# Current Trends on Monomial and Binomial Ideals

# Current Trends on Monomial and Binomial Ideals

Special Issue Editors

**Huy Tài Hà**
**Takayuki Hibi**

MDPI • Basel • Beijing • Wuhan • Barcelona • Belgrade • Manchester • Tokyo • Cluj • Tianjin

*Special Issue Editors*
Huy Tài Hà
Department of Mathematics,
Tulane University
USA

Takayuki Hibi
Department of Pure and Applied
Mathematics, Graduate School
of Information Science and
Technology, Osaka University
Japan

*Editorial Office*
MDPI
St. Alban-Anlage 66
4052 Basel, Switzerland

This is a reprint of articles from the Special Issue published online in the open access journal *Mathematics* (ISSN 2227-7390) (available at: https://www.mdpi.com/journal/mathematics/special issues/Current_Trends_Monomial_Binomial_Ideals).

For citation purposes, cite each article independently as indicated on the article page online and as indicated below:

LastName, A.A.; LastName, B.B.; LastName, C.C. Article Title. *Journal Name* **Year**, *Article Number*, Page Range.

**ISBN 978-3-03928-360-6 (Pbk)**
**ISBN 978-3-03928-361-3 (PDF)**

© 2020 by the authors. Articles in this book are Open Access and distributed under the Creative Commons Attribution (CC BY) license, which allows users to download, copy and build upon published articles, as long as the author and publisher are properly credited, which ensures maximum dissemination and a wider impact of our publications.

The book as a whole is distributed by MDPI under the terms and conditions of the Creative Commons license CC BY-NC-ND.

# Contents

**About the Special Issue Editors** . . . . . . . . . . . . . . . . . . . . . . . . . . . . . . . . . . . . . . . . **vii**

**Preface to "Current Trends on Monomial and Binomial Ideals"** . . . . . . . . . . . . . . . . . . . . **ix**

**Arindam Banerjee and Vivek Mukundan**
Cohen Macaulay Bipartite Graphs and Regular Element on the Powers of Bipartite Edge Ideals
Reprinted from: *Mathematics* **2019**, *7*, 762, doi:10.3390/math7080762 . . . . . . . . . . . . . . . . . . **1**

**Daniel Bănaru and Viviana Ene**
Compatible Algebras with Straightening Laws on Distributive Lattices
Reprinted from: *Mathematics* **2019**, *7*, 671, doi:10.3390/math7080671 . . . . . . . . . . . . . . . . . . **15**

**Huy Tài Hà and Susan Morey**
Algebraic Algorithms for Even Circuits in Graphs
Reprinted from: *Mathematics* **2019**, *7*, 859, doi:10.3390/math7090859 . . . . . . . . . . . . . . . . . . **21**

**Jürgen Herzog and Takayuki Hibi**
The Regularity of Edge Rings and Matching Numbers
Reprinted from: *Mathematics* **2020**, *8*, 39, doi:10.3390/math8010039 . . . . . . . . . . . . . . . . . . . **39**

**Takayuki Hibi and Akiyoshi Tsuchiya**
Odd Cycles and Hilbert Functions of Their Toric Rings
Reprinted from: *Mathematics* **2020**, *8*, 22, doi:10.3390/math8010022 . . . . . . . . . . . . . . . . . . . **43**

**Do Trong Hoang, Giancarlo Rinaldo and Naoki Terai**
Cohen-Macaulay and ($S_2$) Properties of the Second Power of Squarefree Monomial Ideals
Reprinted from: *Mathematics* **2019**, *7*, 684, doi:10.3390/math7080684 . . . . . . . . . . . . . . . . . . **47**

**A. V. Jayanthan and Neeraj Kumar**
Syzygies, Betti Numbers, and Regularity of Cover Ideals of Certain Multipartite Graphs
Reprinted from: *Mathematics* **2019**, *7*, 869, doi:10.3390/math7090869 . . . . . . . . . . . . . . . . . . **55**

**Lukas Katthän**
Linear Maps in Minimal Free Resolutions of Stanley-Reisner Rings
Reprinted from: *Mathematics* **2019**, *7*, 605, doi:10.3390/math7070605 . . . . . . . . . . . . . . . . . . **73**

**Kazunori Matsuda, Hidefumi Ohsugi and Kazuki Shibata**
Toric Rings and Ideals of Stable Set Polytopes
Reprinted from: *Mathematics* **2019**, *7*, 613, doi:10.3390/math7070613 . . . . . . . . . . . . . . . . . . **79**

**Aki Mori**
Faces of 2-Dimensional Simplex of Order and Chain Polytopes
Reprinted from: *Mathematics* **2019**, *7*, 851, doi:10.3390/math7090851 . . . . . . . . . . . . . . . . . . **91**

**S. A. Seyed Fakhari**
On the Stanley Depth of Powers of Monomial Ideals
Reprinted from: *Mathematics* **2019**, *7*, 607, doi:10.3390/math7070607 . . . . . . . . . . . . . . . . . . **97**

**Miguel Eduardo Uribe-Paczk and Adam Van Tuyl**
The Regularity of Some Families of Circulant Graphs
Reprinted from: *Mathematics* **2019**, *7*, 657, doi:10.3390/math7070657 . . . . . . . . . . . . . . . . . . **115**

# About the Special Issue Editors

**Huy Tài Hà** (Professor) obtained a B.Sc. degree in Mathematics and Computing Science in December, 1995, and a B.Sc. (Hons) degree in Mathematics in June, 1996, from Curtin University, Australia. He continued his graduate study at Queen's University, Canada, and received a Ph.D. degree in Mathematics in July 2000. His Ph.D. thesis is entitled "Rational Surfaces from an Algebraic Perspective". From 2000 to 2001, he was a regular research member at the Institute of Mathematics, Hanoi, Vietnam. From 2001 to 2004, he was a postdoctoral fellow at the University of Missouri, USA. In July 2004, he started a tenure-track Assistant Professor position at Tulane University, USA, and has been there since. He was promoted to an Associate Professor with tenure at Tulane University in 2009. He became a Full Professor at Tulane University in 2017. Since December 2015, he has been an editor for the Journal of Algebra and Its Applications. During his academic career, he has written more than 50 papers for peer-reviewed journals and conference proceedings. He is one of the authors of the monograph entitled "Ideals of Powers and Powers of Ideals", which is going to be published by Springer. His research has been supported by the National Science Foundation, the National Security Agency, Simons Foundation and Louisiana Board of Regents.

**Takayuki Hibi** (Professor) studied mathematics at Nagoya University, Japan, and graduated in March 1981 with a Bachelor of Science degree. He received his Ph.D. in December 1987 from Nagoya University for the thesis entitled "Canonical ideals of Cohen-Macaulay partially ordered sets." From April 1985 to September 1990, he was a Research Assistant at Nagoya University. He was visiting the Massachusetts Institute of Technology from August 1988 to July 1989. In October 1990, he moved to Hokkaido University and had been there as Associate Professor until March 1995. Since April 1992, he has been an editor of the Journal of Pure and Applied Algebra. In April 1995, he became a Professor of Mathematics, Osaka University and since then has been in his present office. During his academic career, he has written more than 200 papers for journals and refereed conference proceedings and has been the first supervisor of 10 completed Ph.D. theses and over 100 master's and bachelor's theses. From October 2008 to March 2019, he had been awarded huge grants. His big project on Gröbner bases was concluded with great success. He is one of the authors of the monographs "Monomial Ideals" (GTM 260, Springer) and "Binomial Ideals" (GTM 279, Springer).

# Preface to "Current Trends on Monomial and Binomial Ideals"

Historically, the study of monomial ideals became fashionable after the pioneering work by Richard Stanley in 1975 on the upper bound conjecture for spheres. On the other hand, since the early 1990s, under the strong influence of Gröbner bases, binomial ideals became gradually fashionable in commutative algebra. The last ten years have seen a surge of research work in the study of monomial and binomial ideals. Remarkable developments in, for example, finite free resolutions, syzygies, Hilbert functions, toric rings, as well as cohomological invariants of ordinary powers, and symbolic powers of monomial and binomial ideals, have been brought forward. The theory of monomial and binomial ideals has many benefits from combinatorics and Göbner bases. Simultaneously, monomial and binomial ideals have created new and exciting aspects of combinatorics and Göbner bases.

In the present Special Issue, particular attention was paid to monomial and binomial ideals arising from combinatorial objects including finite graphs, simplicial complexes, lattice polytopes, and finite partially ordered sets, because there is a rich and intimate relationship between algebraic properties and invariants of these classes of ideals and the combinatorial structures of their combinatorial counterparts. This volume gives a brief summary of recent achievements in this area of research. It will stimulate further research that encourages breakthroughs in the theory of monomial and binomial ideals. This volume provides graduate students with fundamental materials in this research area. Furthermore, it will help researchers find exciting activities and avenues for further exploration of monomial and binomial ideals.

The editors express our thanks to the contributors to the Special Issue. Funds for APC (article processing charge) were partially supported by JSPS (Japan Society for the Promotion of Science) Grants-in-Aid for Scientific Research (S) entitled "The Birth of Modern Trends on Commutative Algebra and Convex Polytopes with Statistical and Computational Strategies" (JP 26220701). The publication of this volume is one of the main activities of the grant.

**Huy Tài Hà, Takayuki Hibi**
*Special Issue Editors*

*Article*

# Cohen Macaulay Bipartite Graphs and Regular Element on the Powers of Bipartite Edge Ideals

**Arindam Banerjee [1],* and Vivek Mukundan [2]**

1. Ramakrishna Mission Vivekananda Educational and Research Institute, Belur, West Bengal 711202, India
2. Department of Mathematics, University of Virginia, Charlottesville, VA 22902, USA
* Correspondence: 123.arindam@gmail.com

Received: 8 August 2019; Accepted: 13 August 2019; Published: 20 August 2019

**Abstract:** In this article, we discuss new characterizations of Cohen-Macaulay bipartite edge ideals. For arbitrary bipartite edge ideals $I(G)$, we also discuss methods to recognize regular elements on $I(G)^s$ for all $s \geq 1$ in terms of the combinatorics of the graph $G$.

**Keywords:** Cohen Macaulay; Bipartite graphs; regular elements on powers of bipartite graphs; colon ideals; depth of powers of bipartite graphs; dstab; associated graded rings

## 1. Introduction

The interplay between the combinatorics of finite simple graphs $G$ and the algebra of the underlying edge ideals $I(G)$ has been studied by various researches during the last few decades. The algebraic invariants that have been particularly prone to combinatorial interpretation are regularity, projective dimension, depth, and Betti numbers. In this article, we study the depth of powers of edge ideals of bipartite graphs. Combinatorics of bipartite graphs have been particularly ripe with interesting algebraic counterparts in the edge ideals and their powers. Interested readers are referred to [1–3], etc. In this paper, we continue the study pursued by the same authors in [3]. We study the closely related topics of combinatorial characterization of regular elements and Cohen-Macaulayness of various powers of bipartite edge ideals.

In section two of this paper, we offer a new characterization of Cohen-Macaulay bipartite edge ideals. We characterize it using colon ideals of the form $(I(G)^2 : e)$, where $e$ is an edge/generator of $I(G)$, somehow in the same way as it is done in [3,4], etc., in the study of regularity. An often quoted and important characterization of Cohen-Macaulay bipartite edge ideals is due to Herzog-Hibi in [2]. In this article, we also give a new proof of this characterization ([2]). One important feature of our proof is that it does not use Hall's marriage theorem or any variant of it as it was done in [2]. Throughout this article, we refer to $S$ as the polynomial ring $\mathbf{k}[x_1, \ldots, x_n, y_1, \ldots, y_{n'}]$. Our main results in this section are as follows:

**Theorem 1.** *Let $G$ be a bipartite graph with partition $V_1 = \{x_1, \ldots, x_n\}$ and $V_2 = \{y_1, \ldots, y_{n'}\}$. Then the following are equivalent*

1. *$S/I(G)$ is Cohen-Macaulay*
2. *$n = n'$ and there exists a re-ordering of the vertex sets $V_1, V_2$ such that*

    (a) *$x_i y_i \in I(G)$ for all $i$*
    (b) *If $x_i y_j \in I(G)$ then $i \leq j$.*
    (c) *If $x_i y_j, x_j y_k \in I(G)$ then $x_i y_k \in E$.*

3. *$I(G)$ is unmixed and $S/I(G)$ is connected in codimension one.*
4. *$n = n'$ and there exists exactly $n$ edges $e_1, \ldots, e_n$ such that $(I(G)^2 : e_i) = I(G)$ and for $i \neq j$, $e_i$ and $e_j$ are disjoint.*

5. $n = n'$ and there exists exactly $n$ edges $e_1, \ldots, e_n$ such that $(I(G)^2 : e_i)$ is Cohen-Macaulay and for $i \neq j$, $e_i$ and $e_j$ are disjoint.

For arbitrary bipartite edge ideals, it is often hard to compute the depth of powers of its edge ideals $I(G)^s$ for all $s \geq 1$. Even if $G$ is Cohen-Macaulay, it is not so easy to compute the depth $S/I(G)^s$ for $s \geq 2$. It is well known that depth $S/I(G)^s$ is asymptotically equal to the number of connected components of $G$ ([5]). An important invariant related to the study of depth $S/I(G)^s$ is the dstab $I(G)$ which measures the minimal $t$ for which depth $S/I(G)^t$ equals the number of connected components of $G$. To study such invariants the same authors in [3] characterized regular elements on $I(G)^s$ for any unmixed bipartite graphs $G$. In the third section of this paper we characterize elements of the form $x_\nu - y_\mu$ that are regular on the powers $I(G)^s$ of a bipartite edge ideal $G$. This is a generalization of the similar result proved in [3]. Our characterization turns out to be the exactly same as the $\star$-condition proved there. To signify its usefulness we call it the neighborhood properties (we refer to the definition in Definition 12) Our main result proved here is as follows:

**Theorem 2.** *Let $G$ be a bipartite graph and suppose that $x_\mu \in V_1$ and $y_\nu \in V_2$ satisfies the neighborhood properties. Then $x_\mu - y_\nu$ is an regular element on $S/I(G)^s$ for all $s$.*

## 2. Structure of Cohen-Macaulay and Unmixed Bipartite Graphs

A characterization theorem for Cohen-Macaulay bipartite graphs was given by Herzog-Hibi in [2].

**Theorem 3.** *(Herzog-Hibi, [2]) Let $G$ be a bipartite graph with partition $V_1 = \{x_1, \ldots, x_n\}$ and $V_2 = \{y_1, \ldots, y_{n'}\}$. Then the following are equivalent*

1. *$S/I(G)$ is Cohen-Macaulay*
2. *$n = n'$ and there exists a re-ordering of the vertex sets $V_1, V_2$ such that*

   (a) *$x_i y_i \in I(G)$ for all $i$*
   (b) *If $x_i y_j \in I(G)$ then $i \leq j$.*
   (c) *If $x_i y_j, x_j y_k \in I(G)$ then $x_i y_k \in E$.*

The following theorem is an improvement of the Herzog-Hibi characterization (Theorem 3). We are grateful to Prof. Huneke for the ideas presented in this proof. It is important to notice here that the following theorem does not make use of the Halls marriage theorem which is an important element of any proofs known to us of Theorem 3.

**Definition 1.** *(Definition, p. 498, [6]) Let $I$ be an ideal in a polynomial ring $S$ such that $I = P_1 \cap \cdots \cap P_k$, $P_i \in \mathrm{Spec}(S), 1 \leq i \leq k$. We say that the ring $S/I$ is connected in codimension one if for any two primes $Q', Q'' \in \mathrm{Min}(S/I)$, there is a sequence of minimal primes $Q' = Q_1, \ldots, Q_r = Q'' \in \mathrm{Min}(S/I)$ such that for each $i = 1, 2, \ldots, r-1$, $\mathrm{ht}(Q_i + Q_{i+1}) = 1$ in $S/I$.*

**Theorem 4.** *Let $G$ be a bipartite graph with partition $V_1 = \{x_1, \ldots, x_n\}$ and $V_2 = \{y_1, \ldots, y_{n'}\}$. Then the following are equivalent*

1. *$S/I(G)$ is Cohen-Macaulay*
2. *$n = n'$ and there exists a re-ordering of the vertex sets $V_1, V_2$ such that*

   (a) *$x_i y_i \in I(G)$ for all $i$*
   (b) *If $x_i y_j \in I(G)$ then $i \leq j$.*
   (c) *If $x_i y_j, x_j y_k \in I(G)$ then $x_i y_k \in E$.*
3. *$I(G)$ is unmixed and $S/I(G)$ is connected in codimension one.*

**Proof.** First we show (2) $\Rightarrow$ (1). We prove by induction on $n$. If $n = 1$, then $I(G) = (x_1 y_1)$ and hence clearly $S/I(G)$ is Cohen-Macaulay. Now assume that the result is true for $n-1$ and let $G$ be a graph which satisfies the conditions $(a) - (c)$ of (2) on $2n$ vertices (with partition $V_1 = \{x_1, \ldots, x_n\}$ and $V_2 = \{y_1, \ldots, y_n\}$). Consider

$$0 \to \frac{S}{(I(G) : x_1)} \to \frac{S}{I(G)} \to \frac{S}{((I(G), x_1))} \to 0 \tag{1}$$

Notice that $(I(G), x_1) = (I(G'), x_1)$, where $G'$ is the graph obtained by deleting $x_1$ and $y_1$ from $G$. Clearly $G'$ satisfies the conditions $(a) - (c)$ of (2) and hence $S/I(G')$ is Cohen-Macaulay (on $2n - 2$ vertices) by induction. So $S/(I(G), x_1)$ is Cohen-Macaulay of dimension $n$. Let $\{y_1, y_{i_1}, \ldots, y_{i_k}\} \subseteq (I(G) : x_1)$ for some $i_1, \ldots, i_k$. Let $x_{i_j} y_l \in I(G)$ for some $1 \leq j \leq k$. As $x_1 y_{i_j} \in I(G)$ by the condition $(c)$, $x_1 y_l \in I(G)$ and hence $l \in \{1, i_1, \ldots, i_k\}$. So $(I(G) : x_1) = (I(G''), y_1, \ldots, y_{i_k})$, where $G''$ is the graph obtained from $G$ by deleting $x_1, y_1, x_{i_2}, y_{i_2}, \ldots, x_{i_k}, y_{i_k}$. But by induction, $S/I(G')$ is Cohen-Macaulay of dimension $n - k$. Hence $S/(I(G) : x_1)$ is Cohen-Macaulay of dimension $n$. Now in (1), both $S/(I(G) : x_1)$ and $S/(I(G), x_1)$ are Cohen-Macaulay of dimension $n$, we have $S/I(G)$ is also Cohen-Macaulay of dimension $n$ ((Proposition 1.2.9, [7]) and the fact that dimension of $S/I(G)$ is the maximum of the dimensions of $S/(I(G) : x_1)$ and $S/(I(G), x_1)$).

The implication (1) $\Rightarrow$ (3) is a consequence of (Corollary 2.4, [6]).

We finally show (3) $\Rightarrow$ (2). We first observe that $n = n'$ as $I(G)$ is unmixed and both $(x_1, \ldots, x_n)$ and $(y_1, \ldots, y_{n'})$ are minimal primes. Next, we prove that the existence of conditions $(a)$ and $(b)$ by induction. Let $\emptyset \neq L \subset \{1, \ldots, n\}$ and define

$$y^L = \prod_{i \in L} y_i \qquad x^L = \prod_{i \in L} x_i \qquad T_L = \{j \mid x_j y_i \notin I(G) \text{ for any } i \in L\} \qquad u^L = y^L x^{T_L}.$$

Note that $u^L \notin I(G)$ for any subset $S \subseteq \{1, \ldots, n\}$. We now consider the ideals $(I(G) : u^L)$. If $L' = \{1, \ldots, n\}$ then $(I(G) : u^{L'}) = (x_1, \ldots, x_n)$ which shows that $(x_1, \ldots, x_n) \in \mathrm{Ass}(I(G))$. Since $I(G)$ is unmixed, we have $\mathrm{ht}\, I(G) = n$. Clearly for any $L \subseteq \{1, \ldots, n\}$, $(I(G) : u^L) = (x_{j_1}, \ldots, x_{j_t}, y_{l_1}, \ldots, y_{l_{t'}})$ where for each $1 \leq i \leq t$, $x_{j_i} y_{r_i} \in I(G)$ for some $r_i \in S$ and for each $1 \leq k \leq t'$, $x_{w_k} y_{l_k} \in I(G)$ for some $w_k \in T_s$. Since $I(G)$ is unmixed of height $n$ and $(x_{j_1}, \ldots, x_{j_t}, y_{l_1}, \ldots, y_{l_{t'}}) \in \mathrm{Ass}(I(G))$, we have $t + t' = n$.

Now choose $y_i$ with minimum vertex degree. Without loss of generality we may assume $i = 1$. Let $x_1, \ldots, x_t$ be neighbors of $y_1$ and $L = \{1\}$. Then as in the previous paragraph, consider $(I(G) : u^L) = (x_1, \ldots, x_t, y_{l_1}, \ldots, y_{l_{n-t}})$. After relabeling, we may assume $y_1, \ldots, y_t$ are only connected to $x_1, \ldots, x_t$. Let $G'$ be the induced subgraph on $x_1, \ldots, x_t, y_1, \ldots, y_t$. By our choice of $y_1$, of minimal vertex degree $t$, notice that every other vertex $y_j$ has to have vertex degree at least $t$. In other words, since $t$ is minimal, each vertex $y_i, 1 \leq i \leq t$ in $G'$ has at least $t$ neighbors and hence $G'$ is a complete bipartite graph.

Since $S/I(G)$ is connected in codimension one and $(x_1, \ldots, x_n), (y_1, \ldots, y_n) \in \mathrm{Ass}(I(G))$, there exists a sequence of minimal primes $(x_1, \ldots, x_n) = P_1, \ldots, P_r = (y_1, \ldots, y_n)$ such that $\mathrm{ht}(P_i + P_{i+1}) = 1$ in $S/I(G)$. If any minimal prime $P_l$ of $I(G)$ does not contain some $x_i, 1 \leq i \leq t$ then it has to contain every $y_j, 1 \leq j \leq t$ (as $G'$, as defined in the previous paragraph, is a complete bipartite graph). Let $1 \leq l \leq r$ such that for all $1 \leq i \leq l$, $P_i$ contains all of $x_1, \ldots, x_t$ (alternatively, $P_i$'s do not contain any of $y_1, \ldots, y_t$). Now $P_{l+1}$ does not contain at least one of $x_1, \ldots, x_t$, hence it has to contain all $y_1, \ldots, y_t$. So $\mathrm{ht}(P_l + P_{l+1}) \geq t$ in $S/I(G)$. Thus $t = 1$ and hence $y_1$ is only connected to $x_1$.

Now consider $(I(G), x_1)$. Since $I(G)$ is an intersection of minimal primes, $(I(G), x_1)$ is an intersection of minimal primes of $I(G)$ containing $x_1$. Thus any minimal prime of $(I(G), x_1)$ is a minimal prime of $I(G)$, and so $(I(G), x_1)$ is unmixed. We now show that $(I(G), x_1)$ is connected at codimension one. Any minimal prime of $I(G)$ has to contain either $x_1$ or $y_1$ (as it is minimal it cannot contain both as $y_1$ is only connected to $x_1$). Let $P', P'' \in \mathrm{Min}(I(G), x_1)$. As $P', P'' \in \mathrm{Min}\, I(G)$, there exists a sequence of minimal primes $P' = P_1, \ldots, P_r = P''$ such that ht $P_i + P_{i+1} = 1$. For any $1 \leq i \leq r$,

$$P'_i = \begin{cases} P_i & \text{if } x_1 \in P_i \\ (P_i \backslash \{y_1\}) \cup \{x_1\} & \text{if } x_1 \notin P_i \end{cases}.$$

The sequence $P' = P'_1, \ldots, P'_r = P''$ defined as before has the property that ht $P'_i + P'_{i+1} = 1$ and hence $(I(G), x_1)$ is connected in codimension one. Now notice that $(I(G), x_1) = (I(G''), x_1)$ where $G''$ is the graph obtained from $G$ by deleting $x_1$. By induction hypothesis, $S/I(G'')$ is Cohen-Macaulay. So there exists an ordering $\{x_2, \ldots, x_n\}$ and $\{y_2, \ldots, y_n\}$ satisfying $(a) - (b)$ of $(2)$. As $y_1$ is only connected to $x_1$, $G$ also satisfies $(a) - (b)$ of $(2)$.

To prove that condition $(c)$ holds, take $x_i y_j$ and $x_j y_k$ in $E(G)$ such that $i, j, k$ are distinct. Assume that $x_i y_k$ is not an edge. Then there is a minimal prime $P$ that does not contain either $x_i$ or $y_k$ as the ideal generated by all $x$-variables except $x_i$ and all $y$-variables except $y_k$ is a prime ideal that contains $I(G)$ and does not contain $x_i$ or $y_k$. Now because $I(G)$ is unmixed, height of this prime has to be $n$. Since $x_i$ and $y_k$ are not in $P$, we get that $y_j$ and $x_j$ are both in $P$. As $P$ contains at least one of $x_m$ or $y_m$ for all $m$, one observes that height of $P$ is strictly bigger than $n$, which is a contradiction. □

The following remark is extremely crucial for our work.

**Remark 1.** *If $G$ is a bipartite graph and $ab$ is an edge then from (Theorem 6.7, [4]) we get $(I(G)^2 : ab)) = I(G) + (uv|u \in N(a), v \in N(b))$.*

**Theorem 5.** *Let $G$ be a bipartite graph with partition $V_1 = \{x_1, \ldots, x_n\}$ and $V_2 = \{y_1, \ldots, y_{n'}\}$. Then the following are equivalent*

1. $S/I(G)$ is Cohen-Macaulay
2. $n = n'$ and there exists exactly n edges $e_1, \ldots, e_n$ such that $(I(G)^2 : e_i) = I(G)$ and for $i \neq j$, $e_i$ and $e_j$ are disjoint.
3. $n = n'$ and there exists exactly n edges $e_1, \ldots, e_n$ such that $S/(I(G)^2 : e_i)$ is Cohen-Macaulay and for $i \neq j$, $e_i$ and $e_j$ are disjoint.

**Proof.** First, we show $(1) \Leftrightarrow (2)$. If $S/I(G)$ is Cohen-Macaulay, we have ordering $x_1, \ldots, x_n$ and $y_1, \ldots, y_n$ of the vertices of $G$ which satisfies the conditions of Theorem 4. Condition $(c)$ implies for all $i$, $I(G)^2 : x_i y_i = I(G)$ and conditions $(a)$ and $(b)$ implies for $i \neq j$ $(I(G)^2 : x_i y_j) \neq I(G)$.

Now suppose there exist, after possible reordering, $e_1 = x_1 y_1, \ldots, e_n = x_n y_n$ which satisfied the conditions of $(2)$. First, we show that if $G_i$ is the induced subgraph obtained by deleting $x_i$ and $y_i$ then the edge ideal $J_i$ related to $G_i$ satisfies the condition with $e_1, \ldots, e_{i-1}, e_{i+1}, \ldots, e_n$. Without loss of generality, we prove this for $G_1$. Clearly $(J_1^2 : e_i) = J_1$ for $2 \leq i \leq n$. Suppose there exists an edge $x_i y_j, i \neq j$ such that $(J_1^2 : x_i y_j) = J_1$. Without loss of generality we may assume $i = 2, j = 3$. As $(I(G)^2 : x_2 y_3) \neq I(G)$ and $x_1 y_1$ is an edge we can conclude that there exists a minimal generator of $(I(G)^2 : x_2 y_3)$ which is an edge that is either of the form $x_1 y_l$ or $x_m y_1$ (Theorem 6.7, [4]). Again without loss of generality we may assume it is of the form $x_1 y_l$ as the proof for the other follows simply by interchanging roles of $x$ and $y$. So $x_1 y_3$ and $x_2 y_l$ are edges in $G$ (Theorem 6.7, [4]). As $(J_1^2 : x_2 y_3) = J_1$ we conclude $x_3 y_2$ is an edge in $G$. As $(I(G)^2 : x_3 y_3) = I(G)$ we observe that $x_1 y_2$ has to be an edge in $G$. So $l \neq 2,3$. Without loss of generality we may assume $l = 4$. Now $(I(G)^2 : x_2 y_2) = I(G)$ so $x_3 y_4$ has to be an edge in $G$. Again $(I(G)^2 : x_3 y_3) = I(G)$ hence $x_1 y_4$ is an edge in $G$ contradicting the assumption. So we may assume for all $i$ the edge ideal $I(G_i)$ of the graph $G_i$ obtained by deleting $x_i$ and $y_i$ satisfies the conditions in $(2)$.

Now by induction we may assume the result holds for $n - 1$. Pick $e_i = x_i y_i$ such that $y_i$ has minimum degree. Let $G'$ be the induced subgraph on vertices other than $x_i, y_i$ with edge ideal $I(G')$. As $I(G')$ satisfies the condition it is Cohen-Macaulay by induction. Without loss of generality we may assume $i = 1$ and ordering that gives ordering of previous theorem for $I(G')$ is $x_2, ..., x_n, y_2, ..., y_n$. As $y_2$ has degree one in $G'$ it can have at most degree 2 in $G$. If $x_1y_2$ is not an edge, due to minimality degree of $y_1$ is at most 1. If $x_1y_2$ is an edge in $G$ and $x_iy_1$ is an edge in $G$ for $i > 2$, as $(I(G)^2 : x_1y_1) = I(G)$, we have $x_iy_2$ is an edge in $G$ and hence in $G'$ contradicting the assumption. Now if $x_1y_2$ and $x_2y_1$ both are edges in $G$. Notice that $x_2y_1$ also satisfies the hypothesis $(I(G)^2 : x_2y_1 = I(G))$. For, $x_1$ has to be connected to any neighbor of $x_2$ as $x_1y_2$ is an edge and $x_2y_2$ satisfies the hypothesis $(I(G)^2 : x_2y_2 = I(G))$. This leads to a contradiction and hence no $x_i$ for $i > 1$ is connected to $y_1$. This guarantees that conditions (a) and (b) of Theorem 4(2) is satisfied. The condition (c) is satisfied as for all $i$, $(I(G)^2 : x_iy_i) = I(G)$.

Next we show (1) $\Leftrightarrow$ (3). To prove the if part, we pick, without loss of generality, $y_1$ with minimum degree and the corresponding edge $e_1 = x_1y_1$. If degree of $y_1$ more than one then degree of any other vertex is more than one; as $(I(G)^2 : e_1)$ is Cohen-Macaulay this will be a contradiction to the fact that any Cohen-Macaulay bipartite graph should have a $y$-vertex of degree 1 (Theorem 3). So $y_1$ has degree one. Hence $(I(G)^2 : e_1) = I(G)$ and $I(G)$ is Cohen-Macaulay.

For the only if part let $e_1 = (x_1y_1), ..., e_n = (x_ny_n)$ be as the ordering prescribed by the Herzog-Hibi (Theorem 3) characterization. All we need to show is that $J = (I(G)^2 : x_iy_j)$ is not Cohen-Macaulay for $i > j$. This follows as $(J^2 : e) = J$ for $e = x_jy_i$ (which is a minimal monomial generator of $J$) as well as for $e_1, ..., e_n$. To see this first we show that $(J^2 : e_k) = J$ for all $k$. Here at every step we use the description of colon ideal provided by (Theorem 6.7, [4]). If $x_ly_m$ is a minimal monomial generator of $(J^2 : e_k)$ which is not in $J$ then $x_ly_k$ and $x_ky_m$ are in $J$. Both of them cannot belong to $I(G)$ as from $(I(G)^2 : e_k) = I(G)$ that will imply $x_ly_m$ belongs to $I(G)$ and as a result will belong to $J$, contradicting the assumption. Without loss of generality assume $x_ky_m$ does not belong to $I(G)$. Then $x_ky_j$ and $x_iy_m$ is in $I(G)$. If $x_ly_k$ does not belong to $I(G)$ then $x_ly_j$ and $x_iy_k$ belong to $I(G)$. If $x_ly_k$ is in $I(G)$ as $x_ky_j$ is in $I(G)$ and $(I(G)^2 : e_k) = I(G)$ we have $x_ly_j$ is in $I(G)$. In either case we have $x_ly_j$ and $x_iy_m$ belong to $I(G)$. Hence $x_ly_m$ belongs to $J$ contradicting our assumption.

Next we show that $(J^2 : x_jy_i) = J$. If $x_ly_k$ is a minimal monomial generator of $(J^2 : x_jy_i)$ which is not in $J$ then $x_jy_k$ and $x_ly_i$ is in $J$. As $x_jy_k$ is in $J$ it is either in $I(G)$ or $y_k$ is a neighbor of $x_i$ in $G$. If $x_jy_k$ is in $I(G)$ as $(I(G)^2 : x_jy_j) = I(G)$ we have $x_iy_k$ is in $I(G)$. By symmetry $x_ly_j$ is in $I(G)$. Hence $x_ly_k$ is in $J$ contrary to the assumption. Hence $J$ is not Cohen-Macaulay. □

The next theorem gives insight into the associated graded ring of a Cohen-Macaulay bipartite edge ideal. The proof of this theorem uses the description of the colon of the $n$th power of an edge ideal with $n - 1$ edges introduced in [4].

**Theorem 6.** *Let $I(G)$ be Cohen-Macaulay bipartite edge ideal with an ordering of vertices satisfying Theorem 3(2) and $e_i = x_iy_i$ for $1 \leq i \leq n$. Then for all $i$ and for all $k$, $(I(G)^k : e_i) = I(G)^{k-1}$. Hence $e_i$s are non zero divisors in the associated graded ring of $I(G)$.*

**Proof.** Let $f \in (I(G)^k : e_i) \subset (I(G)^{k-1} : e_i)$ be a minimal monomial generator of $(I(G)^k : e_i)$. By induction $(I(G)^{k-1} : e_i) = I(G)^{k-2}$. So $f = gh_1....h_{k-2}$ where $h_j$s are minimal monomial generators of $I(G)$ and $g$ any monomial. So $e_ih_1....h_{k-2}g \in I(G)^k$. As $f$ is a minimal monomial generator, without loss of generality we may assume $g$ is of degree 2 and $e_ih_1..h_{k-2}g$ is a minimal monomial generator of $I(G)^k$. Let $g = x_ky_l, k \leq l$. If $g$ is an edge we are done. Otherwise by ([4], Theorem 6.7), $x_k$ and $y_l$ are even connected with respect to $e_ih_1...h_{k-2}$. If $x_iy_l$ is an edge and for some $j, m, p, h_j = x_my_p$ and $x_my_i$ is an edge. Then by Theorem 4(2(c)) $x_my_l$ is an edge and hence proceeding inductively we show $g$ is an edge and the result follows. □

We illustrate this theorem for $k = 3, 4$.

**Example 1.** *Let $S = \mathbf{k}[x_1, x_2, x_3, y_1, y_2, y_3]$ and $I = (x_1y_1, x_2y_2, x_3y_3, x_1y_2, x_1y_3, x_2y_3)$. One can check using Macaulay 2, that $(I^3 : x_1y_1) = (I^3 : x_2y_2) = (I^3 : x_3y_3) = I^2$ and $(I^4 : x_1y_1) = (I^4 : x_2y_2) = (I^4 : x_3y_3) = I^3$.*

## 3. Regular Elements in Powers of Bipartite Edge Ideals

This section presents methods to recognize regular elements on the power of bipartite edge ideals based on the combinatorics of the graph. We first present some examples to motivate the definition and the results.

**Example 2.** *Consider the ring* $S = k[x_1, x_2, x_3, y_1, y_2, y_3]$ *and the bipartite edge ideal* $I(G) = (x_1y_1, x_2y_2, x_3y_3, x_1y_2, x_1y_3, x_2y_3)$ *corresponding to*

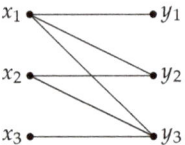

*Macaulay2 computations show that* $x_3 - y_1$ *is a regular element on* $I(G)^s$ *for* $1 \leq s \leq 10$. *Notice that* $I(G)$ *is Cohen-Macaulay. This can also be recovered from (Theorem 3.8, [3]).*

One would be tempted to generalize that $x_n - y_1$ is always a regular element for bipartite graphs. But it is not always the case as it is shown in this example.

**Example 3.** *Consider the ring* $S = k[x_1, x_2, x_3, y_1, y_2, y_3]$ *and the bipartite edge ideal* $I(G) = (x_1y_1, x_2y_2, x_3y_3, x_1y_2, x_2y_1, x_2y_3)$ *corresponding to*

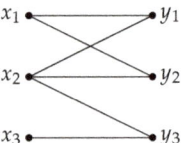

*Macaulay2 computations show that* $x_3 - y_1$ *is not a regular element* $S/I(G)$ *or* $S/I(G)^2$.

Studying more such examples, we came up with the following definition involving the combinatorial nature of the graphs.

**Definition 2.** *Let $G$ be a bipartite graph. Then $x_\mu \in V_1, y_\nu \in V_2$ satisfies the neighborhood condition if*

$$N(x_\mu) \subseteq N(x_{a_i}) \text{ for all } i, 1 \leq i \leq p \text{ where } N(y_\nu) = \{x_{a_1}, \ldots, x_{a_p}\}. \tag{2}$$

**Remark 2.** *Condition (2) of Definition 2 is equivalent to the following condition*

$$N(y_\nu) \subseteq N(y_{b_j}) \text{ for all } i, 1 \leq j \leq q \text{ where } N(x_\mu) = \{y_{b_1}, \ldots, y_{b_q}\}.$$

Suppose (2) of Definition 2 is true. Then $\{y_{b_1}, \ldots, y_{b_q}\} = N(x_\mu) \subseteq N(x_{a_i})$, where $N(y_\nu) = \{x_{a_1}, \ldots, x_{a_p}\}$. This means $x_{a_i} \in N(y_{b_j})$ where $1 \leq i \leq p, 1 \leq j \leq q$. In other words, $N(y_\nu) \subseteq N(y_{b_j})$, where $1 \leq j \leq q$. The other direction is analogous.

We show in [3] that $x_n - y_1$ is a regular element on $S/I(G)^s$ for all $s \geq 1$ when $G$ is an unmixed bipartite graph. Of course, when $G$ is unmixed bipartite, $x_n$ and $y_1$ satisfies the neighborhood conditions. In this section, we show that the difference of vertices which satisfies the neighborhood condition are the right candidates for being a regular element on $S/I(G)^s$ for any bipartite graph $G$.

Theorem 8 is the main theorem we study in this section. We break up the proof of this theorem into three main parts, where Theorem 7, Lemma 1 provide all the tools required to prove Theorem 8.

**Theorem 7.** *Let G be a bipartite graph and suppose that $x_\mu \in V_1$ and $y_\nu \in V_2$ satisfies the neighborhood properties. If m is a monomial such that $mx_\mu^k, my_\nu^k \in I(G)^s$, then $m \in I(G)^s$ for $s, k \geq 1$.*

**Proof.** We prove by induction on $k$. Suppose $k = 1$. Then $mx_\mu, my_\nu \in I(G)^s$. As $mx_\mu \in I(G)^s$, then either $m \in I(G)^s$ or $m = m'y_t$ for some $y_t \in N_G(x_\mu)$ and $m' \in I(G)^{s-1}$. If $m \in I(G)^s$, then the claim is obviously true.

Suppose $m = m'y_t$ with $m' \in I(G)^{s-1} \setminus I(G)^s$. Let $m' = ae_1 \cdots e_{s-1}$ for $e_1, \ldots, e_k \in I(G), a \in S$. We assume $e_i = (x_{u_i}y_{v_i})$, $1 \leq i \leq s-1$. Since $my_\nu \in I(G)^s$, we have $m'y_ty_\nu \in I(G)^s$. Thus

$$m'y_ty_\nu = ae_1 \cdots e_{s-1}y_ty_\nu \in I(G)^s \qquad (3)$$
$$= bf_1 \ldots f_s \text{ for } f_1, \ldots, f_s \in I(G), b \in S$$

Suppose a neighbor of $y_t$ divides $a$, then clearly $m = m'y_t \in I(G)^s$. Now suppose a neighbor of $y_\nu$ divides $a$. Since $x_\mu$ and $y_\nu$ satisfies the neighborhood properties, any neighbor of $y_\nu$ is also a neighbor of $y_t$ and hence $m = m'y_t \in I(G)^s$.

Suppose that no neighbor of $y_t$ or $y_\nu$ divide $a$. Now in the decomposition in (3), if $y_\nu$ does not divide $f_1 \cdots f_s$, then $f_1 \cdots f_s$ divides $m'y_t = m$ and hence $m \in I(G)^s$. Now if $y_t$ does not divide $f_1 \cdots f_s$, then $f_1 \cdots f_s$ divides $m'y_\nu$. Thus $m'y_\nu = b_1 f_1 \cdots f_s$. If $y_\nu$ divides $b_1$, then $f_1 \cdots f_s$ divides $m'$ and hence $m \in I(G)^s$. Now suppose $y_\nu$ divides, say $f_1 = (x_\delta y_\nu)$. Again, since $x_\mu$ and $y_\nu$ satisfy the neighborhood properties, any neighbor of $y_\nu$ is a neighbor of $y_t$ and hence $m = m'y_t = b_1(x_\delta y_t)f_2 \cdots f_s \in I(G)^s$.

Now suppose that $y_ty_\nu$ divides $f_1 \cdots f_s$. Since $y_ty_\nu$ divides $f_1 \cdots f_s$, we assume, without loss of generality, $f_1 = x_{u_1}y_t$ and $f_2 = (x_{u_2}y_\nu)$. Thus we have

$$m'y_ty_\nu = af_1f_2e_3 \cdots e_{s-1}y_{v_1}y_{v_2} \in I(G)^s \qquad (4)$$
$$= bf_1 \ldots f_s \text{ for } f_1, \ldots, f_s \in I(G), b \in S$$

Now a neighbor of $y_{v_1}$, say $x_0$, divides $a$, then

$$m = m'y_t = a'(x_0y_{v_1})(x_{u_1}y_t)e_2 \cdots e_{s-1} \in I(G)^s \text{ where } a = a'x_0$$

Similarly if a neighbor of $y_{v_2}$, say $x_0$, divides $a$, then

$$m = m'y_t = a''(x_0y_{v_2})e_1(x_{u_2}y_t)e_2 \cdots e_{s-1} \in I(G)^s \text{ where } a = a''x_0$$

Now suppose no neighbor of $y_{v_1}$ or $y_{v_2}$ divides $a$. Consider (4). If $y_{v_1}$ does not divide $f_3 \cdots f_s$, then

$$m'y_ty_\nu = af_1f_2e_3 \cdots e_{s-1}y_{v_1}y_{v_2}$$
$$= b'y_{v_1}f_1 \ldots f_s = b'y_{v_1}(x_{u_1}y_t)(x_{u_2}y_\nu)f_3 \cdots f_s$$
$$= b'(x_{u_1}y_{v_1})(x_{u_2}y_t)f_3 \cdots f_sy_\nu$$
$$= b'e_1(x_{u_2}y_t)f_3 \cdots f_sy_\nu$$

Deleting $y_\nu$ on both sides, we get $m = m'y_t = b'(x_{u_1}y_{v_1})(x_{u_2}y_t)f_3 \cdots f_s \in I(G)^s$. Thus we assume $y_{v_1}$ divides $f_3 \cdots f_s$ and hence assume, without loss of generality $f_3 = (x_{u_3}y_{v_1})$. Now we have

$$m'y_ty_\nu = af_1f_2f_3e_4 \cdots e_{s-1}y_{v_2}y_{v_3} \in I(G)^s \qquad (5)$$
$$= bf_1 \ldots f_s \text{ for } f_1, \ldots, f_s \in I(G), b \in S$$

Now if $y_{v_2}$ does not divides $f_4 \cdots f_s$, then

$$\begin{aligned}
m'y_ty_v &= af_1f_2f_3e_4\cdots e_{s-1}y_{v_2}y_{v_3}\\
&= b'y_{v_2}f_1\ldots f_s = b'y_{v_2}(x_{u_1}y_t)(x_{u_2}y_v)f_3\cdots f_s\\
&= b'(x_{u_1}y_t)(x_{u_2}y_{v_2})f_3\cdots f_sy_v\\
&= b'f_1e_2f_3\cdots f_sy_v
\end{aligned}$$

Deleting $y_v$ on both sides we get $m = m'y_t = b'f_1e_2f_3\cdots f_s \in I(G)^s$.

Thus we assume $y_{v_2}$ divide $f_4 \cdots f_s$ and hence assume, without loss of generality, $f_4 = (x_{u_4}y_{v_2})$. We now have

$$\begin{aligned}
m'y_ty_v &= af_1f_2f_3f_4e_5\cdots e_{s-1}y_{v_3}y_{v_4} \in I(G)^s\\
&= bf_1\ldots f_s \text{ for } f_1,\ldots,f_s \in I(G), b \in S
\end{aligned} \tag{6}$$

We continue in the same fashion and arrive at the $j$-th decomposition

$$\begin{aligned}
m'y_ty_v &= af_1\cdots f_{2j-1}f_{2j}e_{2j+1}\cdots e_{s-1}y_{v_{2j-1}}y_{v_{2j}} \in I(G)^s\\
&= bf_1\ldots f_s \text{ for } f_1,\ldots,f_s \in I(G), b \in S
\end{aligned} \tag{7}$$

Also $f_{2r-1} = (x_{u_{2r-1}}y_{v_{2r-3}})$ and $f_{2r} = (x_{u_{2r}}y_{v_{2r-2}})$ for $2 \leq r \leq j$. Now if a neighbor of $y_{v_{2j-1}}$, say $x_0$, divides $a$, then

$$m = m'y_t = a'(x_0y_{v_{2j-1}})f_1e_2f_3e_4\cdots e_{2j-2}f_{2j-1}e_{2j}e_{2j+1}\cdots e_{s-1} \in I(G)^s \tag{8}$$
$$\text{where } a = a'x_0$$

If a neighbor of $y_{v_{2j}}$, say $(x_0)$, divides $a$, then

$$m = m'y_t = a'(x_0y_{v_{2j}})e_1(x_{u_2}y_t)e_3f_4e_5f_6\cdots e_{2j-1}f_{2j}e_{2j+1}e_{2j+2}\cdots e_{s-1} \in I(G)^s \tag{9}$$
$$\text{where } a = a'x_0$$

Now suppose no neighbor of $y_{v_{2j-1}}$ or $y_{v_{2j}}$ divides $a$. Now consider (7). If $y_{v_{2j-1}}$ does not divide $f_{2j+1}\cdots f_s$, then

$$\begin{aligned}
m'y_ty_v &= af_1\cdots f_{2j-1}f_{2j}e_{2j+1}\cdots e_{s-1}y_{v_{2j-1}}y_{v_{2j}}\\
&= b'y_{v_{2j-1}}f_1\ldots f_s\\
&= b'e_1(x_{u_2}y_t)e_3f_4e_5f_6\cdots f_{2j-2}e_{2j-1}f_{2j}f_{2j+2}\cdots f_sy_v
\end{aligned} \tag{10}$$

Deleting $y_v$ on both sides we have $m = m'y_t = b'e_1(x_{u_2}y_t)e_3f_4e_5f_6\cdots f_{2j-2}e_{2j-1}f_{2j}f_{2j+2}\cdots f_s \in I(G)^s$. Thus we assume $y_{v_{2j-1}}$ divides $f_{2j+1}\cdots f_s$ and hence assume, without loss of generality, $f_{2j+1} = (x_{u_{2j+1}}y_{v_{2j-1}})$. We now have

$$\begin{aligned}
m'y_ty_v &= af_1\cdots f_{2j-1}f_{2j}f_{2j+1}e_{2j+2}\cdots e_{s-1}y_{v_{2j-1}}y_{v_{2j}} \in I(G)^s\\
&= bf_1\ldots f_s \text{ for } f_1,\ldots,f_s \in I(G), b \in S
\end{aligned} \tag{11}$$

Again, if $y_{v_{2j}}$ does not divide $f_{2j+2}\cdots f_s$, then

$$m'y_t y_v = a f_1 \cdots f_{2j-1} f_{2j} f_{2j+1} e_{2j+2} \cdots e_{s-1} y_{v_{2j-1}} y_{v_{2j}} \quad (12)$$
$$= b' y_{v_{2j}} f_1 \cdots f_s$$
$$= b' f_1 e_2 f_3 e_4 f_5 \cdots f_{2j-1} e_{2j} f_{2j+2} f_{2j+2} \cdots f_s y_v$$

Deleting $y_v$ on both sides, we get $m = m' y_t = b' f_1 e_2 f_3 e_4 f_5 \cdots f_{2j-1} e_{2j} f_{2j+2} f_{2j+2} \cdots f_s \in I(G)^s$. Thus we assume $y_{v_{2j}}$ divides $f_{2j+2} \cdots f_s$ and hence assume, without loss of generality, $f_{2j+2} = (x_{u_{2j+2}} y_{v_{2j}})$.

Continuing in the same fashion we may reach the final decomposition

$$m' y_t y_v = a f_1 \cdots f_{s-1} y_{v_{s-2}} y_{v_{s-1}} \in I(G)^s \quad (13)$$
$$= b f_1 \ldots f_s \text{ for } f_1, \ldots, f_s \in I(G), b \in S$$

Recall that every stage we make sure that none of the neighbors of the $y$'s appearing in $f_1, \ldots, f_{s-1}$ divide $a$. Thus a neighbor of $y_{v_{s-2}}$ or $y_{v_{s-1}}$ divides $a$. Now we can use the decomposition in (8) and (9) to show that $m \in I(G)^s$ depending on whether $s-2$ or $s-1$ is odd or even. This concludes the proof of claim of this theorem in $k = 1$ case.

Now assume by induction, that if $m x_\mu^l, m y_v^l \in I(G)^s$ for $1 \leq l \leq k-1$, then $m \in I(G)^s$. Suppose $m x_\mu^k, m y_v^k \in I(G)^s$. We also assume that $k \leq s$. For, if $k > s$, then $m x \mu^s, m y_v^s \in I(G)^s$ and hence by induction hypothesis, we have $m \in I(G)^s$.

We claim that it is enough to show that $m x_\mu^{k-1} \in I(G)^s$ or $m y_v^{k-1} \in I(G)^s$. Suppose we show that $m x_\mu^{k-1} \in I(G)^s$. We now have $m x_\mu^{k-1}, m y_v^k \in I(G)^s$. Thus $m x_\mu^{k-1} y_v, m y_v^{k-1} y_v \in I(G)^s$ and hence $(m y_v) x_\mu^{k-1}, (m y_v) y_v^{k-1} \in I(G)^s$. Since $m y_v$ is a monomial, we use induction hypothesis to conclude that $m y_v \in I(G)^s$. Thus we now have $m x_\mu^{k-1}, m y_v \in I(G)^s$. As before, we have $(m x_\mu^{k-2}) x_\mu, (m x_\mu^{k-2}) y_v \in I(G)^s$. Again, since $m x_\mu^{k-2}$ is a monomial, we use induction hypothesis to conclude that $m x_\mu^{k-2} \in I(G)^s$. We now have $m x_\mu^{k-2}, m y_v \in I(G)^s$. We continue the process to get $m x_\mu^2, m y_v \in I(G)^s$. We still have $(m x_\mu) x_\mu, (m x_\mu) y_v \in I(G)^s$. Since $m x_\mu$ is a monomial, by induction hypothesis, we get $m x_\mu \in I(G)^s$. We now have $m x_\mu, m y_v \in I(G)^s$. This is the $k=1$ case. We now use the induction hypothesis to get $m \in I(G)^s$. On the other hand, if we show that $m y_v^{k-1} \in I(G)^s$, then we can analogously show that $m \in I(G)^s$.

Now we go to the induction step. We have $m x_\mu^k, m y_v^k \in I(G)^s$. Since $m x_\mu^k \in I(G)^s$ and $m x_\mu^l \notin I(G)^s$ for any $l < k$, we have $m = m' y_{t_1} \cdots y_{t_k}$ where $m' \in I(G)^{s-k}, y_{t_1}, \ldots, y_{t_k} \in N_G(x_\mu)$ and not all $y_{t_1}, \ldots, y_{t_k}$ may be distinct. Suppose a neighbor of $y_{t_1}, \ldots, y_{t_k}$ divides $a$, then $m x_\mu^{k-1} \in I(G)^s$.

Now suppose no neighbor of $y_{t_1}, \ldots, y_{t_k}$ divide $a$. Since $m y_v^k \in I(G)^s$ we have

$$m y_v^k = m' y_{t_1} \cdots y_{t_k} y_v^k \in I(G)^s \quad (14)$$
$$= b f_1 \cdots f_s \text{ where } f_1, \ldots, f_s \in I(G), b \in S$$

We observe that $m'$ may be written divisible by many minimal monomial generators of $I(G)^{s-k}$. We can take $m' = a e_1 \ldots e_{s-k}$ such that $\frac{m'}{e_1 \ldots e_{s-k}}$ has smallest number of $x$ variables in common with $f_1 \ldots f_s$.

It is clear that $y_v^k$ must divide $f_1 \ldots f_s$, otherwise $m y_v^l \in I(G)^s$ for some $l < k$ and hence $m y_v^{k-1} \in I(G)^s$. Recall the no neighbor of $y_{t_1}, \ldots, y_{t_k}$ divides $a$. Thus we can assume that no neighbor of $y_v$, divides $a$ as that will make $m x_\mu^{k-1} \in I(G)^s$. So without loss of generality we may assume for $1 \leq i \leq k, f_i = x_{u_i} y_v$ where for every $j, e_j = x_{u_j} y_{v_j}$.

Now we observe that if any neighbor of $y_{v_i}$ for $1 \leq i \leq k$ divide $a$ then, clearly, $m x_\mu^{k-1} \in I(G)^s$. For, without loss of generality, say $x_0 y_{v_1}$ is an edge where $x_0$ divides $a$. As $x_{u_1} y_v$ is an edge, so is $x_{u_1} y_{t_1}$ (by neighborhood properties). Thus we have $m = (\frac{a}{x_0})(x_0 y_{v_1}) e_2 \ldots e_{s-k}(x_{u_1} y_{t_1}) \ldots y_{t_k} \in I(G)^{s-k+1}$. Hence this will force $m x_\mu^{k-1} \in I(G)^s$. So we assume no neighbor of $y_{v_i}$ for $1 \leq i \leq k$ divide $a$.

As there are $s$ many $x$ variables in $f_1 \cdots f_s$ and $k < s$, some of the $x$ variables of $f_1 \cdots f_s$ divides $a$. We also have that no neighbor of any $y_{t_i}$ divides $a$ and $y_v^k$ divides $f_1 \cdots f_s$. Let $f_{k+1} = x_0 y_{v_{k+1}}$ where $x_0$ divides $a$ and $e_{k+1} = x_{u_{k+1}} y_{v_{k+1}}$. We may write $m' = a' e_1 \ldots e_k f_{k+1} e_{k+2} \ldots e_{s-k}$ where $a' = (\frac{a}{x_0} x_{u_{k+1}})$. But this is an expression of $m'$ with $a'$ having less number of $x$ variables in common with $f_1 \ldots f_s$ than $a$ which is a contradiction. Thus, one of the neighbors of $y_{v_i}$ for some $1 \leq i \leq k$ divides $a$ and hence $m \in I(G)^s$. □

**Lemma 1.** *Let $G$ be a bipartite graph and suppose that $x_\mu \in V_1$ and $y_\nu \in V_2$ satisfies the neighborhood properties. Now assume $m_1, \ldots, m_k \in S$ are monomials of the same degree such that $(m_1 + \cdots + m_k)(x_\mu - y_\nu) \in I(G)^s$. Further suppose,*

$$m_1 x_\mu = m_2 y_\nu \tag{15}$$

$$m_i x_\mu = m_{i+1} y_\nu \text{ for } 2 \leq i \leq k-1 \tag{16}$$

$$m_1 y_\nu, m_k x_\mu \in I(G)^s \tag{17}$$

*Then $m_j \in I(G)^s$ for $1 \leq j \leq k$.*

**Proof.** First, assume that $N_G(y_\nu) = \{x_{v_1}, \ldots, x_{v_p}\}$. We prove by induction on $k$. If $k = 1$, then clearly the claim is true by Theorem 7. By induction, assume the claim is true for $(m_1 + \cdots + m_l)(x_\mu - y_\nu) \in I(G)^s$ satisfying (15)–(17) and $l \leq k - 1$. Now suppose we have

$$(m_1 + \cdots + m_k)(x_\mu - y_\nu) \in I(G)^s$$

satisfying (15)–(17). We show that $m_1 \in I(G)^s$. This will show that $(m_2 + \cdots + m_k)(x_\mu - y_\nu) \in I(G)^s$ satisfying (15)–(17). Thus by induction hypothesis we have $m_j \in I(G)^s$ for $2 \leq j \leq k$ proving the claim.

From (15), we have $m_1 = m y_\nu$ and $m_2 = m x_\mu$ where $m \in S$, a monomial. From (16), we have $m_3 = \frac{m_2 x_\mu}{y_\nu} = \frac{m x_\mu^2}{y_\nu}$. Subsequently, we show that

$$m_i = \frac{m x_\mu^{i-1}}{y_\nu^{i-2}} \text{ for } 2 \leq i \leq k \tag{18}$$

Since $m_1 y_\nu \in I(G)^s$, we have $m_1 \in I(G)^s$ or $m_1 = a e_1 \cdots e_{s-1} x_{v_t}$ for some $t \in \{1, \ldots, p\}$ where $N_G(y_\nu) = \{x_{v_1}, \ldots, x_{v_p}\}$.

Suppose $m_1 = a e_1 \cdots e_{s-1} x_{v_1}$. Since $m_1 = m y_\nu$, $y_\nu$ divides $a$ or one of the $e_i$'s. If $y_\nu$ divides $a$, then $m_1 \in I(G)^s$.

Now suppose $y_\nu$ divides, say $e_1 = x_{v_b} y_\nu$ for some $b \in \{1, \ldots, p\}$. Since $m_1 = m y_\nu$, we have $m = a e_2 \cdots e_{s-1} x_{v_1} x_{v_b}$. Using this equality in (18), we have

$$m_k = \frac{m x_\mu^{k-1}}{y_\nu^{k-2}} = \frac{a e_2 \cdots e_{s-1} x_{v_1} x_{v_b} x_\mu^{k-1}}{y_\nu^{k-2}}$$

Since $y_\nu$ does not divide $a$, then $y_\nu$ divides some of the $e_1, \ldots, e_{s-k}$ and hence we have $k - 2 \leq s - 2$ or $k \leq s$. Without loss of generality, assume $y_\nu$ divides $e_2, \ldots, e_{k-1}$. Thus

$$m_k = a e_k \cdots e_{s-1} x_{v_1}^{l_1} \cdots x_{v_p}^{l_p} x_\mu^{k-1} \text{ where } \sum_{j=1}^{p} l_j = k$$

Let $u = a e_k \cdots e_{s-1} x_{v_1}^{l_1} \cdots x_{v_p}^{l_p}$. Now as $m_k x_\mu \in I(G)^s$, we have $u x_\mu^k \in I(G)^s$. Also, notice that

$$u y_\nu^k = \frac{m_k}{x_\mu^{k-1}} y_\nu^k = \frac{m}{y_\nu^{k-2}} y_\nu^k = m y_\nu^2 = m_1 y_\nu \in I(G)^s$$

Since $u$ is a monomial, we have $u \in I(G)^s$, by Theorem 7. Now $m_1 = u y_v^{k-1} \in I(G)^s$ and hence we are done. □

We now prove one of the main results of this section. In this theorem, we attempt to rearrange the sum $m_1 + \cdots + m_k$ into a configuration shown in the previous lemma.

**Theorem 8.** *Let $G$ be a bipartite graph and suppose that $x_\mu \in V_1$ and $y_\nu \in V_2$ satisfies the neighborhood properties. Then $x_\mu - y_\nu$ is an regular element on $S/I(G)^s$ for all $s$.*

**Proof.** Consider $(m_1 + \cdots + m_k)(x_\mu - y_\nu) \in I(G)^s$ where $m_i$'s are monomials of the same degree. We prove $m_1, \ldots, m_k \in I(G)^s$ by induction on $k$.

Suppose $k = 1$ and $m_1(x_\mu - y_\nu) = m_1 x_\mu - m_1 y_\nu \in I(G)^s$. Thus $m_1 x_\mu, m_1 y_\nu \in I(G)^s$. Now we use Theorem 7, to show that $m_1 \in I(G)^s$ proving the base case of induction.

Suppose $(m_1 + \cdots + m_l)(x_\mu - y_\nu) \in I(G)^s$ for $l \leq k-1$ implies $m_1, \ldots, m_l \in I(G)^s$. Now consider

$$(m_1 + \cdots + m_k)(x_\mu - y_\nu) = m_1 x_\mu - m_1 y_\nu + m_2 x_\mu - m_2 y_\nu + \cdots + m_k x_\mu - m_k y_\nu \in I(G)^s. \quad (19)$$

where all $m_i$'s are distinct. We show $m_i \in I(G)^s$ for $1 \leq i \leq k$.

Observe that if $m_1 x_\mu, m_1 y_\nu \in I(G)^s$, then we have $m_1(x_\mu - y_\nu) \in I(G)^s$ and $(m_2 + \cdots + m_k)(x_\mu - y_\nu) \in I(G)^s$. Now we use induction hypothesis to show that $m_i \in I(G)^s$ for $1 \leq i \leq k$.

Now we first consider the following configuration, i.e., after possible re-ordering of $m_i$'s we have

$$m_1 x_\mu = m_2 y_\nu \quad (20)$$
$$m_i x_\mu = m_{i+1} y_\nu, \text{ for } 2 \leq i \leq k-1 \quad (21)$$
$$m_k x_\mu = m_1 y_\nu \quad (22)$$

We refer to this case as the *k-cancellation* case. Using (20), we get $m_1 = m y_\nu$ and $m_2 = m x_\mu$. Using this and (21), we get

$$m_i = \frac{m x_\mu^{i-1}}{y_\nu^{i-2}} \text{ for } 3 \leq i \leq k \quad (23)$$

Thus $m_k = \frac{m x_\mu^{k-1}}{y_\nu^{k-2}}$. Using this description in (22) we get $x_\mu^k = y_\nu^k$, a contradiction.

Now consider (19). Without loss of generality, after possible reordering, assume that $m_1 x_\mu = m_2 y_\nu$. If $m_1 y_\nu = m_2 x_\mu$, then we get $(m_1 + m_2)(x_\mu - y_\nu) \in I(G)^s$ and $(m_3 + \cdots + m_k)(x_\mu - y_\nu) \in I(G)^s$. Now using induction hypothesis, we get $m_i \in I(G)^s$.

Suppose, if $m_1 y_\nu = m_3 x_\mu$ we introduce the re-ordering

$$m_1^{(1)} = m_3, m_2^{(1)} = m_1, m_3^{(1)} = m_2$$
$$m_i^{(1)} = m_i \text{ for } 4 \leq i \leq k-1$$

Notice that $(m_1 + \cdots + m_k)(x_\mu - y_\nu) = \left(m_1^{(1)} + \cdots + m_k^{(1)}\right)(x_\mu - y_\nu)$. Thus it is enough to show that $m_i^{(1)} \in I(G)^s$. Under this re-ordering $m_1^{(1)} x_\mu = m_2^{(1)} y_\nu$ and $m_2^{(1)} x_\mu = m_3^{(1)} y_\nu$. If $m_1^{(1)} y_\nu = m_3^{(1)} x_\mu$, then we get $(m_1^{(1)} + m_2^{(1)} + m_3^{(1)})(x_\mu - y_\nu) \in I(G)^s$ and $(m_4^{(1)} + \cdots + m_k^{(1)})(x_\mu - y_\nu) \in I(G)^s$. Now using induction hypothesis, we get $m_i^{(1)} \in I(G)^s$ and hence $m_i \in I(G)^s$.

Now if $m_1^{(1)} y_v = m_4^{(1)} x_\mu$, we introduce a new ordering

$$m_1^{(2)} = m_4^{(1)}$$
$$m_l^{(2)} = m_{l-1}^{(1)} \text{ for } 2 \le l \le 4$$
$$m_q^{(2)} = m_q^{(1)} \text{ for } 5 \le q \le k$$

As before we consider if $m_1^{(2)} y_v = m_4^{(2)} x_\mu$ or $m_1^{(2)} y_v = m_5^{(2)} x_\mu$ and introduce new ordering, if necessary.

We now continue this process and arrive at the $j$-th re-ordering defined as follows

$$m_1^{(j)} = m_{j+2}^{(j-1)}$$
$$m_l^{(j)} = m_{l-1}^{(j-1)} \text{ for } 2 \le l \le j+2$$
$$m_q^{(j)} = m_q^{(j-1)} \text{ for } j+3 \le q \le k$$

with the following configuration

$$m_i^{(j)} x_\mu = m_{i+1}^{(j)} y_v \text{ for } 1 \le i \le j+1$$

First, suppose $j = k-2$. As before, we consider two cases $m_1^{(j)} y_v = m_{j+2}^{(j)} x_\mu$ or $m_1^{(j)} y_v \ne m_{j+2}^{(j)} x_\mu$. If $m_1^{(j)} y_v = m_{j+2}^{(j)} x_\mu$, then we arrive at the $k$-cancellation case discussed above, which leads to a contradiction. So we have $m_1^{(j)} y_v \ne m_{j+2}^{(j)} x_\mu$ which is discussed separately in Lemma 1, showing that $m_i \in I(G)^s$.

Now we assume $j < k-2$ and $m_1^{(j)} y_v \ne m_t^{(j)} x_\mu$ for $2 \le t \le k$. If $m_{j+2}^{(j)} x_\mu \ne m_t^{(j)} y_v$ for $j+3 \le t \le k$, then we have $(m_1^{(j)} + \cdots + m_{j+2}^{(j)})(x_\mu - y_v) \in I(G)^s$ and $(m_{j+2}^{(j)} + \cdots + m_k^{(j)})(x_\mu - y_v) \in I(G)^s$ and we use induction hypothesis to conclude that $m_i^{(j)} \in I(G)^s$ and hence $m_i \in I(G)^s$ for $1 \le i \le k$.

Thus assume $m_{j+2}^{(j)} x_\mu = m_t^{(j)} y_v$ for some $j+3 \le t \le k$. Now we use the ordering

$$m_{j+3}^{(j,1)} = m_t^{(j)}, m_t^{(j,1)} = m_{j+3}^{(j)}$$
$$m_i^{(j,1)} = m_i^{(j)} \text{ for } i \ne j+3, t$$

with the configuration $m_i^{(j,1)} x_\mu = m_{i+1}^{(j,1)} y_v$ for $1 \le i \le j+2$.

Now if $m_{j+3}^{(j,1)} x_\mu \ne m_a^{(j,1)} y_v$ for $j+4 \le a \le k$, then $(m_1^{(j,1)} + \cdots + m_{j+3}^{(j,1)})(x_\mu - y_v) \in I(G)^s$ and $(m_{j+4}^{(j,1)} + \cdots + m_k^{(j,1)})(x_\mu - y_v) \in I(G)^s$ and we use induction hypothesis to conclude that $m_i^{(j,1)} \in I(G)^s$ and hence $m_i \in I(G)^s$ for $1 \le i \le k$.

Now if $m_{j+3}^{(j,1)} x_\mu = m_a^{(j,1)} y_v$ for some $j+4 \le a \le k$, then we use the ordering as before

$$m_{j+4}^{(j,2)} = m_a^{(j,1)}, m_a^{(j,2)} = m_{j+4}^{(j,1)}$$
$$m_i^{(j,2)} = m_i^{(j,1)} \text{ for } i \ne j+4, a$$

with the configuration $m_i^{(j,2)} x_\mu = m_{i+1}^{(j,2)} y_v$ for $1 \le i \le j+3$.

We continue in the same fashion to reach $(j,l)$-th re-ordering to get

$$(m_1^{(j,l)} + \cdots m_k^{(j,l)})(x_\mu - y_v) \in I(G)^s$$

with the following configuration

$$m_i^{(j,l)} x_\mu = m_{i+1}^{(j,l)} y_\nu \text{ for } 1 \leq i \leq j+l+1$$

Suppose $j + l = k - 2$, then $m_1^{(j,l)} y_\nu, m_k^{(j,l)} x_\mu \in I(G)^s$. Now using Lemma 1 we have $m_i^{(j,l)} \in I(G)^s$, $1 \leq i \leq k$ and hence $m_i \in I(G)^s$ for $1 \leq i \leq k$.

If $j + l < k - 2$, then there exists a term $m_b^{(j,l)}$ such that $m_b^{(j,l)}(x_\mu - y_\nu) \in I(G)^s$ and $\left( \sum_{t \neq b} m_t^{(j,l)} \right)(x_\mu - y_\nu) \in I(G)^s$ and hence we are done by induction. □

**Corollary 1.** *Let $G$ be a bipartite graph. Suppose $x_\mu \in V_1, y_\nu \in V_2$. Then $x_\mu$ and $y_\nu$ satisfies the neighborhood properties, if and only if $x_\mu - y_\nu$ is regular on $S/I(G)^s$ for all $s$.*

**Proof.** Suppose $x_\mu$ and $y_\nu$ satisfies the neighborhood properties, then $x_\mu - y_\nu$ is regular on $S/I(G)^s$ for all $s$ by Theorem 8.

Now if $x_\mu$ and $y_\nu$ does not satisfy the neighborhood properties, then there exists $y_p$ such that $x_\mu y_p \in E(G)$ and $x_{\nu 1} y_p \notin E(G)$ where $x_{\nu 1} \in N(y_\nu)$. Thus for all $s$ and $e = x_{\nu 1} y_{\nu 1} \in I(G)$,

$$e^{s-1}(x_{\nu 1} y_p)(x_\mu - y_\nu) = e^{s-1}((x_{\nu 1} y_p) x_\mu - (x_{\nu 1} y_p) y_\nu)$$
$$= e^{s-1}(x_{\nu 1}(y_p x_\mu) - (x_{\nu 1} y_\nu) y_p)$$

Since $y_p x_\mu, x_{\nu 1} y_\nu \in I(G)$, we get $e^{s-1}(x_{\nu 1} y_p)(x_\mu - y_\nu) \in I(G)^s$. Thus $x_\mu - y_\nu$ is not a regular element on $I(G)^s$. □

**Author Contributions:** Both the authors contributed equally to the creation of this article.

**Funding:** The first author was partially supported by DST INSPIRE (India) research grant (DST/INSPIRE/04/2017/000752) and he would like to acknowledge that.

**Acknowledgments:** The first author is very thankful to Professor C. Huneke for his valuable suggestions especially regarding the proof of Theorem 5.

**Conflicts of Interest:** The authors declare no conflict of interest.

## References

1. Villarreal, R.H. Unmixed bipartite graphs. *Rev. Colomb. Mat.* **2007**, *41*, 393–395.
2. Herzog, J.; Hibi, T. Distributive lattices, bipartite graphs and Alexander duality. *J. Algebr. Combin.* **2005**, *22*, 289–302. [CrossRef]
3. Banerjee, A.; Mukundan, V. On the Powers of Unmixed Bipartite Edge Ideals. *J. Algebra Appl.* **2018**. [CrossRef]
4. Banerjee, A. The regularity of powers of edge ideals. *J. Algebr. Combin.* **2015**, *41*, 303–321. [CrossRef]
5. Herzog, J.; Hibi, T. Monomial ideals. In *Graduate Texts in Mathematics*; Springer: London, UK, 2011; Volume 260, p. xvi+305.
6. Hartshorne, R. Complete intersections and connectedness. *Amer. J. Math.* **1962**, *84*, 497–508. [CrossRef]
7. Bruns, W.; Herzog, J. Cohen-Macaulay rings. In *Cambridge Studies in Advanced Mathematics*; Cambridge University Press: Cambridge, UK, 1993; Volume 39, p. xii+403.
8. Morey, S.; Reyes, E.; Villarreal, R.H. Cohen-Macaulay, shellable and unmixed clutters with a perfect matching of König type. *J. Pure Appl. Algebra* **2008**, *212*, 1770–1786. [CrossRef]
9. Zaare-Nahandi, R. Cohen-Macaulayness of bipartite graphs, revisited. *Bull. Malays. Math. Sci. Soc.* **2015**, *38*, 1601–1607. [CrossRef]

© 2019 by the authors. Licensee MDPI, Basel, Switzerland. This article is an open access article distributed under the terms and conditions of the Creative Commons Attribution (CC BY) license (http://creativecommons.org/licenses/by/4.0/).

Article

# Compatible Algebras with Straightening Laws on Distributive Lattices

Daniel Bănaru [1] and Viviana Ene [2,*]

[1] Faculty of Mathematics and Computer Science, University of Bucharest, Academiei 14, Bucharest 010014, Romania
[2] Faculty of Mathematics and Computer Science, Ovidius University, Bd. Mamaia 124, Constanta 900527, Romania
* Correspondence: vivian@univ-ovidius.ro

Received: 17 June 2019; Accepted: 26 July 2019; Published: 27 July 2019

**Abstract:** We characterize the finite distributive lattices on which there exists a unique compatible algebra with straightening laws.

**Keywords:** distribuive lattice; algebras with straightening laws; order and chain polytopes

---

## 1. Introduction

Let $P$ be a finite partially ordered set (poset for short) and $\mathcal{I}(P)$ the distributive lattice of the poset ideals of $P$. A subset $\alpha$ of $P$ is a poset ideal of $P$ if it satisfies the following condition: for every $x \in \alpha$ and $y \in P$, if $y \leq x$, then $y \in \alpha$. By a famous theorem of Birkhoff [1], for every finite distributive lattice $L$, there exists a unique subposet $P$ of $L$ such that $L \cong \mathcal{I}(P)$. The order polytope $\mathcal{O}(P)$ and the chain polytope $\mathcal{C}(P)$ were introduced in [2]. In [3], it was shown that the toric ring $K[\mathcal{O}(P)]$ over a field $K$ is an algebra with straightening laws (ASL in brief) on the distributive lattice $\mathcal{I}(P)$ over the field $K$. In [4], it was shown that the ring $K[\mathcal{C}(P)]$ associated with the chain polytope shares the same property.

Let $S = K[x_1, \ldots, x_n, t]$ be the polynomial ring over a field $K$ and $\{w_\alpha\}_{\alpha \in \mathcal{I}(P)}$ be an arbitrary set of monomials in $x_1, \ldots, x_n$ indexed by $\mathcal{I}(P)$. Let $K[\Omega] \subset S$ be the toric ring generated over $K$ by the set of monomials $\Omega = \{\omega_\alpha\}_{\alpha \in \mathcal{I}(P)}$ where $\omega_\alpha = w_\alpha t$ for all $\alpha \in \mathcal{I}(P)$. Clearly, $K[\Omega]$ is a graded algebra if we set $\deg(\omega_\alpha) = 1$ for all $\alpha \in \mathcal{I}(P)$. Let $\varphi : \mathcal{I}(P) \to K[\Omega]$ be the injective map defined by $\varphi(\alpha) = \omega_\alpha$ for all $\alpha \in \mathcal{I}(P)$. Assume that $K[\Omega]$ is an ASL on $\mathcal{I}(P)$ over $K$. According to [4], $K[\Omega]$ is a *compatible* ASL if each of its straightening relations is of the form $\varphi(\alpha)\varphi(\alpha') = \varphi(\beta)\varphi(\beta')$ with $\beta \subseteq \alpha \cap \alpha'$ and $\beta' \supseteq \alpha \cup \alpha'$, where $\alpha, \alpha'$ are incomparable elements in $\mathcal{I}(P)$. If $K[\Omega]$ and $K[\Omega']$ are compatible ASL on $\mathcal{I}(P)$ over $K$, we identify them if they have the same straightening relations. In this case, we write $K[\Omega] \equiv K[\Omega']$.

In ([4], Question 5.1), Hibi and Li asked the following questions:

(a) Given a finite poset $P$, find all possible compatible algebras with straightening laws on $\mathcal{I}(P)$ over $K$.
(b) For which posets $P$, does there exist a unique compatible ASL on $\mathcal{I}(P)$ over $K$?

In this note, we give a complete answer to question (b). Namely, we prove the following:

**Theorem 1.** *Let $P$ be a finite poset. Then, the following statements are equivalent:*

(i) *There exists a unique compatible ASL on $\mathcal{I}(P)$ over $K$.*
(ii) $K[\mathcal{O}(P)] \equiv K[\mathcal{C}(P)] \equiv K[\mathcal{C}(P^*)]$, *where $P^*$ denotes the dual poset of $P$.*
(iii) *Each connected component of $P$ is a chain, that is, $P$ is a direct sum of chains.*

An answer to question (a) seems to be quite difficult. In ([4], Example 5.2), it was observed that, if one considers the poset $P = \{a, b, c, d, e\}$ with $a < c < e$ and $b < c < d$, then there exist nine compatible ASL structures on $\mathcal{I}(P)$ over $K$, while if one considers $P = \{a, b, c, d\}$ with $a < c$, $b < c$, $b < d$, then there are three compatible ASL structures on $\mathcal{I}(P)$ over $K$, namely, $K[\mathcal{O}(P)]$, $K[\mathcal{C}(P)]$, and $K[\mathcal{C}(P^*)]$.

## 2. Order Polytopes, Chain Polytopes, and Their Associated Toric Rings

Let $P = \{p_1, \ldots, p_n\}$ be a finite poset. For the basic terminology regarding posets used in this paper, we refer to [1] and ([5], Chapter 3). The order polytope $\mathcal{O}(P)$ is defined as

$$\mathcal{O}(P) = \{(x_1, \ldots, x_n) \in \mathbb{R}^n : 0 \leq x_i \leq 1, 1 \leq i \leq n, \text{ and } x_i \geq x_j \text{ if } p_i \leq p_j \text{ in } P\}.$$

In ([2], Corollary 1.3), it was shown that the vertices of $\mathcal{O}(P)$ are $\sum_{p_i \in \alpha} \mathbf{e}_i$, $\alpha \in \mathcal{I}(P)$. Here, $\mathbf{e}_i$ denotes the unit coordinate vector in $\mathbb{R}^n$. If $\alpha = \emptyset$, then the corresponding vertex in $\mathcal{O}(P)$ is the origin of $\mathbb{R}^n$.

The chain polytope $\mathcal{C}(P)$ is defined as

$$\mathcal{C}(P) = \{(x_1, \ldots, x_n) \in \mathbb{R}^n : x_i \geq 0, 1 \leq i \leq n,$$

$$x_{i_1} + \cdots + x_{i_r} \leq 1 \text{ if } p_{i_1} < \cdots < p_{i_r} \text{ is a maximal chain in } P\}.$$

In ([2], Theorem 2.2), it was proved that the vertices of $\mathcal{C}(P)$ are $\sum_{p_i \in A} \mathbf{e}_i$, where $A$ is an antichain in $P$. Recall that an antichain in $P$ is a subset of $P$ such that any two distinct elements in the subset are incomparable. Since every poset ideal is uniquely determined by its antichain of maximal elements, it follows that $\mathcal{O}(P)$ and $\mathcal{C}(P)$ have the same number of vertices. However, as it was observed in [2], $\mathcal{O}(P)$ and $\mathcal{C}(P)$ need not have the same number of $i$-dimensional faces for $i > 0$. Therefore, in general, they are not combinatorial equivalent. Combinatorially, equivalence of order and chain polytopes are studied in [6].

*The Toric Rings $K[\mathcal{O}(P)]$ and $K[\mathcal{C}(P)]$*

To each subset $W \subset P$, we attach the squarefree monomial $u_W \in K[x_1, \ldots, x_n]$, $u_W = \prod_{p_i \in W} x_i$. If $W = \emptyset$, then $u_W = 1$. The toric ring $K[\mathcal{O}(P)]$, known as the Hibi ring associated with the distributive lattice $\mathcal{I}(P)$, is generated over $K$ by all the monomials $u_\alpha t \in S$, where $\alpha \in \mathcal{I}(P)$. The toric ring $K[\mathcal{C}(P)]$ is generated by all the monomials $u_A t$ where $A$ is an antichain in $P$. In addition, as we have already mentioned in the Introduction, both rings are algebras with straightening laws on $\mathcal{I}(P)$ over $K$.

We recall the definition of an ASL as it was introduced in [7]. For a quick introduction to this topic, we refer to [7] and ([8], Chapter XIII). Algebras with straightening laws turned out to be useful tools in studying determinantal rings. Let $K$ be a field, $R = \bigoplus_{i \geq 0} R_i$ with $R_0 = K$ be a graded $K$-algebra, $H$ a finite poset, and $\varphi : H \to R$ an injective map which maps each $\alpha \in H$ to a homogeneous element $\varphi(\alpha) \in R$ with $\deg \varphi(\alpha) \geq 1$. A *standard monomial* in $R$ is a product $\varphi(\alpha_1)\varphi(\alpha_2)\cdots\varphi(\alpha_k)$ where $\alpha_1 \leq \alpha_2 \leq \cdots \leq \alpha_k$ in $H$.

**Definition 1.** *The K-algebra R is called an algebra with straightening laws on H over K if the following conditions hold:*

(1) *The set of standard monomials is a $K$–basis of $R$;*
(2) *If $\alpha, \beta \in H$ are incomparable and if $\varphi(\alpha)\varphi(\beta) = \sum c_i \varphi(\gamma_{i1})\ldots\varphi(\gamma_{ik_i})$, where $c_i \in K \setminus \{0\}$ and $\gamma_{i1} \leq \ldots \leq \gamma_{ik_i}$, is the unique expression of $\varphi(\alpha)\varphi(\beta)$ as a linear combination of standard monomials, then $\gamma_{i1} \leq \alpha, \beta$ for all $i$.*

The above relations $\varphi(\alpha)\varphi(\beta) = \sum c_i \varphi(\gamma_{i1})\ldots\varphi(\gamma_{ik_i})$ are called the straightening relations of $R$ and they generate all the relations of of $R$.

Let us go back to the toric rings $K[\mathcal{O}(P)]$ and $K[\mathcal{C}(P)]$.

One considers $\varphi : \mathcal{I}(P) \to K[\mathcal{O}(P)]$ defined by $\varphi(\alpha) = u_\alpha t$ for every $\alpha \in \mathcal{I}(P)$. As it was proved by Hibi in [3], $K[\mathcal{O}(P)]$ is an ASL on $\mathcal{I}(P)$ over $K$ with the straightening relations $\varphi(\alpha)\varphi(\beta) = \varphi(\alpha \cap \beta)\varphi(\alpha \cup \beta)$, where $\alpha, \beta$ are incomparable elements in $\mathcal{I}(P)$.

On the other hand, one defines $\psi : \mathcal{I}(P) \to K[\mathcal{C}(P)]$ by setting $\psi(\alpha) = u_{\max \alpha} t$ for all $\alpha \in \mathcal{I}(P)$ where $\max \alpha$ denotes the set of the maximal elements in $\alpha$. Note that, for every $\alpha \in \mathcal{I}(P)$, $\max \alpha$ is an antichain in $P$ and each antichain $A \subset P$ determines a unique ideal $\alpha \in \mathcal{I}(P)$, namely, the poset ideal generated by $A$. Therefore, $\psi$ is an injective well defined map and by ([4], Theorem 3.1), the ring $K[\mathcal{C}(P)]$ is an ASL on $\mathcal{I}(P)$ over $K$ with the straightening relations

$$\psi(\alpha)\psi(\beta) = \psi(\alpha * \beta)\psi(\alpha \cup \beta),$$

where $\alpha * \beta$ is the poset ideal of $P$ generated by $\max(\alpha \cap \beta) \cap (\max \alpha \cup \max \beta)$.

We observe that one may also consider $K[\mathcal{C}(P^*)]$ as an ASL on $\mathcal{I}(P)$, where $P^*$ is the dual poset of $P$. We may define $\delta : \mathcal{I}(P) \to K[C(P^*)]$ by $\delta(\alpha) = u_{\min \overline{\alpha}} t$ for $\alpha \in \mathcal{I}(P)$, where $\min \overline{\alpha}$ is the set of minimal elements in $\overline{\alpha}$ and $\overline{\alpha}$ is the filter $P \setminus \alpha$ of $P$. We recall that a *filter* $\gamma$ in $P$ (or *dual order ideal*) is a subset of $P$ with the property that for every $p \in \gamma$ and every $q \in P$ with $q \geq p$, we have $q \in \gamma$. Thus, a filter in $P$ is simply a poset ideal in the dual poset $P^*$. The ring $K[C(P^*)]$ is an ASL on $\mathcal{I}(P)$ over $K$ as well with the straightening relations

$$\delta(\alpha)\delta(\beta) = \delta(\alpha \cap \beta)\delta(\alpha \circ \beta)$$

for incomparable elements $\alpha, \beta \in \mathcal{I}(P)$, where $\alpha \circ \beta$ is the poset ideal of $P$ which is the complement in $P$ of the filter generated by $\min(\overline{\alpha} \cap \overline{\beta}) \cap (\min \overline{\alpha} \cup \min \overline{\beta})$. Let us also observe that all the algebras $K[\mathcal{O}(P)], K[\mathcal{C}(P)]$, and $K[\mathcal{C}(P^*)]$ are compatible algebras with straightening laws.

## 3. Proof of Theorem 1

We clearly have (i) $\Rightarrow$ (ii). Let us now prove (ii) $\Rightarrow$ (iii). By hypothesis, the straightening relations of $K[\mathcal{O}(P)], K[\mathcal{C}(P)]$, and $K[\mathcal{C}(P^*)]$ coincide. Therefore, we must have

$$\alpha \cap \beta = \alpha * \beta \text{ and } \overline{\alpha} \cap \overline{\beta} = \overline{\alpha \circ \beta} \qquad (1)$$

for all $\alpha, \beta$ incomparable elements in $\mathcal{I}(P)$. From the second equality in (1), it follows that $\overline{\alpha} \cap \overline{\beta}$ is the filter of $P$ generated by $\min(\overline{\alpha} \cap \overline{\beta}) \cap (\min \overline{\alpha} \cup \min \overline{\beta})$. Assume that there exists two incomparable elements $p, p' \in P$ such that there exists $q \in P$ with $q > p$ and $q > p'$. Consider $\overline{\alpha}$ the filter generated by $p$ and $\overline{\beta}$ the filter generated by $p'$. Then, $\min(\overline{\alpha} \cap \overline{\beta}) \cap (\min \overline{\alpha} \cup \min \overline{\beta}) = \emptyset$, but obviously, $\overline{\alpha} \cap \overline{\beta} \neq \emptyset$. This shows that, for any two incomparable elements $p, p' \in P$, there is no upper bound for $p$ and $p'$.

Similarly, by using the first equality in Equation (1), we derive that, for any two incomparable elements $p, p' \in P$, there is no lower bound for $p$ and $p'$. This shows that every connected component of the poset $P$ is a chain.

Finally, we prove (iii) $\Rightarrow$ (i). Let $P$ be a poset such that all its connected components are chains and assume that the cardinality of $P$ is equal to $n$. Let $\{\omega_\alpha\}_{\alpha \in \mathcal{I}(P)}$ be the generators of $K[\Omega] \subset S$ and assume that the straightening relations of $K[\Omega]$ are $\varphi(\alpha)\varphi(\alpha') = \varphi(\beta)\varphi(\beta')$ where $\beta \subseteq \alpha \cap \alpha', \beta' \supseteq \alpha \cup \alpha'$, and $\alpha, \alpha'$ are incomparable elements in $\mathcal{I}(P)$. We have to show that, for all $\alpha, \alpha'$ incomparable elements in $\mathcal{I}(P)$, we have $\beta = \alpha \cap \alpha'$ and $\beta' = \alpha \cup \alpha'$.

We proceed by induction on

$$k = n - (\text{rank}(\alpha \cup \alpha') - \text{rank}(\alpha \cap \alpha')).$$

Let us recall that, if $\gamma \in \mathcal{I}(P)$, then rank $\gamma$ denotes the rank of the subposet of $\mathcal{I}(P)$ consisting of all elements $\delta \in \mathcal{I}(P)$ with $\delta \subseteq \gamma$.

If $k = 0$, that is, $\text{rank}(\alpha \cup \alpha') - \text{rank}(\alpha \cap \alpha') = n$, then $\alpha \cup \alpha' = P$ and $\alpha \cap \alpha' = \emptyset$, thus $\beta = \alpha \cap \alpha'$ and $\beta' = \alpha \cup \alpha'$. Assume that the desired conclusion is true for $\text{rank}(\alpha \cup \alpha') - \text{rank}(\alpha \cap \alpha') = n - k$ with $k \geq 0$. Let us choose now $\alpha, \alpha'$ incomparable in $\mathcal{I}(P)$ such that $\text{rank}(\alpha \cup \alpha') - \text{rank}(\alpha \cap \alpha') = n - k - 1$ and assume that we have a straightening relation $\varphi(\alpha)\varphi(\alpha') = \varphi(\beta)\varphi(\beta')$ with $\beta \subsetneq \alpha \cap \alpha'$ or $\beta' \supsetneq \alpha \cup \alpha'$. By duality, we may reduce to considering $\beta' \supsetneq \alpha \cup \alpha'$. In other words, in $K[\Omega]$, we have

$$\omega_\alpha \omega_{\alpha'} = \omega_\beta \omega_{\beta'}, \text{ with } \beta \subseteq \alpha \cap \alpha' \text{ and } \beta' \supsetneq \alpha \cup \alpha'.$$

As $P$ is a direct sum of chains, we may find $p \in \max(\alpha \cup \alpha')$ and $q \in \beta' \setminus (\alpha \cup \alpha')$ such that $q$ covers $p$ in $P$, that is, $q > p$ and there is no other element $q'$ in $P$ with $q > q' > p$. Without loss of generality, we may assume that $p \in \alpha'$. Let $\alpha_1$ be the poset ideal of $P$ generated by $\alpha' \cup \{q\}$. As all the connected components of $P$ are chains, we have $\alpha_1 = \alpha' \cup \{q\}$ since there are no other elements in $P$ which are smaller than $q$ except those that are on the same chain as $p$ and $q$, which are in $\alpha'$. Moreover, by the choice of $q$, we have

$$\alpha_1 \subseteq \beta' \text{ and } \alpha \cap \alpha_1 = \alpha \cap (\alpha' \cup \{q\}) = \alpha \cap \alpha'.$$

On the other hand,

$$\text{rank}(\alpha \cup \alpha_1) - \text{rank}(\alpha \cap \alpha_1) = \text{rank}(\alpha \cup \alpha' \cup \{q\}) - \text{rank}(\alpha \cap \alpha')$$

$$= \text{rank}(\alpha \cup \alpha') + 1 - \text{rank}(\alpha \cap \alpha') = n - k.$$

By the inductive hypothesis, it follows that $\varphi(\alpha)\varphi(\alpha_1) = \varphi(\alpha \cap \alpha_1)\varphi(\alpha \cup \alpha_1)$, or, equivalently, in $K[\Omega]$ we have the equality $\omega_\alpha \omega_{\alpha_1} = \omega_{\alpha \cap \alpha_1} \omega_{\alpha \cup \alpha_1}$. Thus, we have obtained the following equalities in $K[\Omega]$ :

$$\omega_\alpha \omega_{\alpha'} = \omega_\beta \omega_{\beta'} \text{ and } \omega_\alpha \omega_{\alpha_1} = \omega_{\alpha \cap \alpha'} \omega_{\alpha \cup \alpha_1}.$$

This implies that

$$\omega_{\alpha \cap \alpha'} \omega_{\alpha'} \omega_{\alpha \cup \alpha_1} = \omega_\beta \omega_{\alpha_1} \omega_{\beta'}. \tag{2}$$

In addition, we have:

$$\alpha \cap \alpha' \subset \alpha' \subset \alpha \cup \alpha_1 \text{ and } \beta \subset \alpha \cap \alpha' = \alpha \cap \alpha_1 \subset \alpha_1 \subset \beta'.$$

This implies that the monomials in Equation (2) are distinct standard monomials in $K[\Omega]$, which is in contradiction to the condition that the standard monomials form a $K$-basis in $K[\Omega]$. Therefore, our proof is completed.

**Author Contributions:** The authors have the same contribution.

**Funding:** This research received no external funding.

**Acknowledgments:** We would like to thank the anonymous referees for their valuable comments.

**Conflicts of Interest:** The authors declare no conflict of interest.

## References

1. Birkhoff, G. *Lattice Theory*, 3rd ed.; American Mathematical Society Colloquium Publications: Providence, RI, USA, 1940; Volume 25.
2. Stanley, R. Two poset polytopes. *Discrete Comput. Geom.* **1986**, *1*, 9–23. [CrossRef]
3. Hibi, T. Distributive lattices, affine semigroup rings and algebras with straightening laws. In *Commutative Algebra and Combinatorics*; Nagata, M., Matsumura, H., Eds.; Advanced Studies in Pure Mathematics: Amsterdam, The Netherlands, 1987; Volume 11, pp. 93–109.
4. Hibi, T.; Li, N. Chain polytopes and algebras with straightening laws. *Acta Math. Vietnam.* **2015**, *40*, 447–452. [CrossRef]
5. Stanley, R. *Enumerative Combinatorics*, 2nd ed.; Cambridge University Press: Cambridge, UK, 2012; Volume I.
6. Hibi, T.; Li, N. Unimodular equivalence of order and chain polytopes. *Math. Scand.* **2016**, *118*, 5–12. [CrossRef]
7. Eisenbud, D. Introduction to algebras with straightening laws, in Ring Theory and Algebra. *Lect. Notes Pure Appl. Math.* **1980**, *55*, 243–268.
8. Hibi, T. *Algebraic Combinatorics on Convex Polytopes*; Carslaw Publications: Glebe, Australia, 1992.

 © 2019 by the authors. Licensee MDPI, Basel, Switzerland. This article is an open access article distributed under the terms and conditions of the Creative Commons Attribution (CC BY) license (http://creativecommons.org/licenses/by/4.0/).

Article
# Algebraic Algorithms for Even Circuits in Graphs

Huy Tài Hà [1] and Susan Morey [2],*

[1] Department of Mathematics, Tulane University, 6823 St. Charles Avenue, New Orleans, LA 70118, USA; tha@tulane.edu
[2] Department of Mathematics, Texas State University, 601 University Drive, San Marcos, TX 78666, USA
\* Correspondence: morey@txstate.eu

Received: 1 July 2019; Accepted: 12 September 2019; Published: 17 September 2019

**Abstract:** We present an algebraic algorithm to detect the existence of and to list all indecomposable even circuits in a given graph. We also discuss an application of our work to the study of directed cycles in digraphs.

**Keywords:** graph; circuit; even cycle; directed cycle; monomial ideal; Rees algebra; edge ideal

**JEL Classification:** 05C38; 13D02; 05C20; 13A30

## 1. Introduction

Detecting the existence of cycles in graphs is a fundamental problem in graph theory (cf. [1–15]). Graph theoretic algorithms exist to enumerate both odd and even cycles. In [16], the first author, together with Francisco and Van Tuyl, gave an algebraic algorithm to detect and exhibit all induced odd cycles in an undirected graph. The work in [16] is an example of the rich interaction between commutative algebra and graph theory. In fact, using algebraic methods to study combinatorial structures and using combinatorial data to understand algebraic properties and invariants has evolved to be an active research topic in combinatorial commutative algebra in recent years (cf. [17,18] and references therein).

In the present paper, we continue this line of work and describe an algebraic algorithm to enumerate even circuits in an undirected graph; a *circuit* is a closed walk in which the edges are all distinct and a *cycle* is a closed walk in which the vertices are all distinct. We also discuss an application of our work to the problem of finding directed cycles in a directed graph (digraph). Let $G = (V, E)$ be a finite simple undirected graph on the vertex set $V = \{x_1, \ldots, x_n\}$. Let $\Bbbk$ be a field and identify the vertices in $V$ with the variables in $R = \Bbbk[x_1, \ldots, x_n]$. The *edge ideal* of $G$ is defined to be

$$I(G) = \langle x_i x_j \mid x_i x_j \in E \rangle.$$

The construction of edge ideals of graphs was first introduced by Villarreal in [19] (see also [20,21]) for edge ideals of simplicial complexes and hypergraphs) and has been an essential tool in various studies in this area of research. Our main result states that even circuits in $G$ can be detected by considering the *reduced Jacobian dual* of the edge ideal $I(G)$, a notion which we define in Section 2.

An even circuit is called *indecomposable* if it cannot be realized as the edge-disjoint union of two smaller even circuits. Our first theorem reads as follows, leaving unexplained terminology until Section 2.

**Theorem 1.** *Let G be a graph, $I = I(G)$, and $\phi$ the presentation matrix from the Taylor resolution of I. Then, the indecomposable even circuits of G correspond exactly to the binomial minors of the reduced Jacobian dual $\overline{B(\phi)}$, which satisfy the following conditions:*

1. the monomials in these binomials are square-free and relatively prime; and
2. the columns of the corresponding submatrices are pairwise center-distinct.

We focus on even circuits because they form a larger class than that of even cycles. With a slight modification of Condition (2), we can also obtain an algebraic characterization for cycles of even lengths in $G$ (see Remark 5). The proof of Theorem 1 is based on an ad hoc analysis of the possible forms of minors of the reduced Jacobian dual $\overline{B(\phi)}$.

Theorem 1 allows us to derive an algebraic algorithm to enumerate all indecomposable even circuits in a given graph that runs in polynomial time on the size of the edge set of the graph (see Algorithm 1). Our goal is not to compare the running time of our algorithm to that of existed ones, rather we aim to exhibit yet another interesting connection between commutative algebra and graph theory. Theorem 1 also has an algebraic consequence to finding defining equations for the Rees algebras of edge ideals of graphs (see Theorem 3).

Theorem 1 furthermore has an interesting application toward the study of directed cycles in digraphs. For a digraph $D$, we construct a bipartite graph $G = G(D)$ (see Definition 7). Note that this bipartite graph has a natural perfect matching, which we denote by $M_D$. There is an established equivalence between the directed cycles in $D$ and the even cycles in $G$ with a certain property that traces its roots back to work done by Dulmage and Mendelsohn in the 1950s (see, for example, [22,23]), which we restate for convenience. Specifically, again leaving unexplained terminology until later, we have:

**Theorem 2.** *The directed cycles in a digraph D correspond exactly to the even cycles in $G = G(D)$ in which a collection of alternating edges forms a subset of the perfect matching $M_D$.*

Theorem 2, combined with Algorithm 1, gives an algebraic algorithm to enumerate all directed cycles in digraphs (see Corollary 4). As a consequence of Theorems 1 and 2, we are also able to translate the famous Caccetta–Häggkvist conjecture for directed cycles in digraphs to a statement about binomial minors of the Jacobian dual matrix (see Conjecture 5).

## 2. Preliminaries

In this section, we collect important notations and definitions used in the paper. For unexplained terminology in commutative algebra, we refer the reader to [24,25], and, in graph theory, we refer the reader to [26].

**Algebra.** Throughout the paper, $\Bbbk$ denotes an infinite field. Let $R = \Bbbk[x_1, \ldots, x_n]$ be a polynomial ring over $k$ and let $\mathfrak{m} = (x_1, \ldots, x_n)$. Let $I \subseteq R$ be an ideal and use $\mu(I)$ to denote the minimal number of generators of $I$. Let $\phi$ be a presentation matrix of $I$.

**Definition 1.** *The* Rees algebra *of $I$ is defined to be the graded ring*

$$\mathcal{R}(I) = R[It] = \mathcal{R} \oplus It \oplus I^2 t^2 \oplus \cdots \subseteq R[t].$$

*Suppose that $I = (f_1, \ldots, f_r)$. Then, there exists a natural presentation of the Rees algebra of I, namely,*

$$R[T_1, \ldots, T_r] \xrightarrow{\theta} R[It] \to 0,$$

*given by $T_i \mapsto f_i t$ for $i = 1, \ldots, r$, where $T_1, \ldots, T_r$ are indeterminates. Set $J = \ker \theta$. Then, $R[It] \cong R[T_1, \ldots, T_r]/J$, and $J$ is referred to as the* ideal of equations *or* defining ideal *of $R[It]$. Since $\phi(T_i) = f_i t$, we say that $T_i$ corresponds to the generator $f_i$ of $I$.*

By the definition of a presentation matrix $\phi$, the linear (in the variables $T_1, \ldots, T_r$) equations of $J$ are generated by entries of the matrix $[T_1 \ \ldots \ T_r] \cdot \phi$. When these entries are linear in $x_1, \ldots, x_n$, that is, when the entries of $\phi$ are linear, then $\phi$ is the Jacobian matrix of these equations with respect to $T_1,\ldots,T_r$. In this setting, one can also define another Jacobian matrix of the same polynomials in $[T_1 \ \ldots \ T_r] \cdot \phi$ but with respect to $x_1,\ldots,x_n$. This new Jacobian matrix is usually denoted by $B(\phi)$ and referred to as the *Jacobian dual* of $\phi$. We now give the generalized version of this notion when the entries of $\phi$ are not necessarily all linear. See [Section 1.5] in [27,28] for further information on Jacobian duals.

**Definition 2** (Jacobian dual). *Let $r = \mu(I)$ and let $\phi$ be a presentation matrix of $I$ with respect to a set of $r$ generators of $I$.*

1. *A Jacobian dual of $\phi$, denoted by $B(\phi)$, is defined to be a matrix, whose entries are in $R[T_1,\ldots,T_r]$ and linear in the variables $T_1,\ldots,T_r$, that satisfies the equation*

$$[T_1 \ldots T_r] \cdot \phi = [x_1 \ldots x_n] \cdot B(\phi).$$

2. *The reduced Jacobian dual of $\phi$, denoted by $\overline{B(\phi)}$, is defined to be $B(\phi) \otimes_k R/\mathfrak{m}$.*

Observe that, given a fixed $\phi$, there may be more than one choice for $B(\phi)$, but $\overline{B(\phi)}$ exists uniquely up to elementary row and column rearrangements that come from re-orderings (see, for example, [28]). The matrix $B(\phi)$, or $\overline{B(\phi)}$, has served as a source for the higher degree generators of $J$ (see, e.g., [28–32]), with the emphasis being on minors of $\overline{B(\phi)}$.

**Example 1.** *Consider the graph*

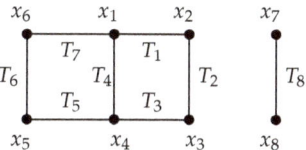

*corresponding to $I = (x_1x_2, x_2x_3, x_3x_4, x_1x_4, x_4x_5, x_5x_6, x_6x_1, x_7x_8)$. Then,*

$$\phi = \begin{bmatrix} x_3 & 0 & 0 & x_4 & 0 & x_6 & 0 & 0 & 0 & x_3x_4 & x_4x_5 & x_5x_6 & x_7x_8 \\ -x_1 & x_4 & 0 & 0 & 0 & 0 & 0 & 0 & 0 & 0 & 0 & 0 & 0 \\ 0 & -x_2 & -x_1 & 0 & -x_5 & 0 & 0 & 0 & 0 & -x_1x_2 & 0 & 0 & 0 \\ 0 & 0 & x_3 & -x_2 & 0 & 0 & -x_5 & 0 & -x_6 & 0 & 0 & 0 & 0 \\ 0 & 0 & 0 & 0 & x_3 & 0 & x_1 & -x_6 & 0 & 0 & -x_1x_2 & 0 & 0 \\ 0 & 0 & 0 & 0 & 0 & 0 & x_4 & x_1 & 0 & 0 & 0 & -x_1x_2 & 0 \\ 0 & 0 & 0 & 0 & 0 & -x_2 & 0 & -x_5 & x_4 & 0 & 0 & 0 & 0 \\ 0 & 0 & 0 & 0 & 0 & 0 & 0 & 0 & 0 & 0 & 0 & 0 & -x_1x_2 \end{bmatrix} \cdots$$

*where the remaining columns of $\phi$ correspond to the rest of the (quadratic) Koszul relations on disjoint pairs of edges. The Koszul relations involving $T_1$ have been included for illustration. Now,*

$$[T_1,\ldots,T_8] \cdot \phi =$$
$$(x_3T_1 - x_1T_2, x_4T_2 - x_2T_3, x_3T_4 - x_1T_3, x_4T_1 - x_2T_4, x_3T_5 - x_5T_3, x_6T_1 - x_2T_7,$$
$$x_1T_5 - x_5T_4, x_4T_6 - x_6T_5, x_1T_6 - x_5T_7, x_4T_7 - x_6T_4, x_3x_4T_1 - x_1x_2T_3,$$
$$x_4x_5T_1 - x_1x_2T_5, x_5x_6T_1 - x_1x_2T_6, x_7x_8T_1 - x_1x_2T_8, \cdots).$$

When using these equations to form $B(\phi)$ as in Definition 2, the nonlinear terms (in the variables $x_i$s) are ambiguous. For example, $x_3x_4T_1$ can be viewed as $x_3(x_4T_1)$ or as $x_4(x_3T_1)$. Different choices of $B(\phi)$ arise from different interpretations for each such nonlinear term in the $x_i$s. The coefficient of $x_i$ of the jth equation goes in the $(i,j)$ entry of $B(\phi)$. One such choice of $B(\phi)$ is

$$B(\phi) = \begin{bmatrix} -T_2 & 0 & -T_3 & 0 & 0 & T_5 & 0 & T_6 & 0 & -x_2T_3 & -x_2T_5 & -x_2T_6 & 0 \\ 0 & -T_3 & 0 & -T_4 & 0 & -T_7 & 0 & 0 & 0 & 0 & 0 & 0 & -x_1T_8 \\ T_1 & 0 & T_4 & 0 & T_5 & 0 & 0 & 0 & 0 & x_4T_1 & 0 & 0 & 0 \\ 0 & T_2 & 0 & T_1 & 0 & 0 & T_6 & 0 & T_7 & 0 & x_5T_1 & 0 & 0 \\ 0 & 0 & 0 & 0 & -T_3 & 0 & -T_4 & 0 & -T_7 & 0 & 0 & x_6T_1 & 0 \\ 0 & 0 & 0 & 0 & 0 & T_1 & 0 & -T_5 & 0 & -T_4 & 0 & 0 & 0 \\ 0 & 0 & 0 & 0 & 0 & 0 & 0 & 0 & 0 & 0 & 0 & 0 & x_8T_1 \\ 0 & 0 & 0 & 0 & 0 & 0 & 0 & 0 & 0 & 0 & 0 & 0 & 0 \end{bmatrix} \cdots$$

*Tensoring with* $R/\mathfrak{m}$ *yields*

$$\overline{B(\phi)} = \begin{bmatrix} -T_2 & 0 & -T_3 & 0 & 0 & 0 & T_5 & 0 & T_6 & 0 & 0 & 0 & 0 & 0 \\ 0 & -T_3 & 0 & -T_4 & 0 & -T_7 & 0 & 0 & 0 & 0 & 0 & 0 & 0 & 0 \\ T_1 & 0 & T_4 & 0 & T_5 & 0 & 0 & 0 & 0 & 0 & 0 & 0 & 0 & 0 \\ 0 & T_2 & 0 & T_1 & 0 & 0 & 0 & T_6 & 0 & T_7 & 0 & 0 & 0 & 0 \\ 0 & 0 & 0 & 0 & -T_3 & 0 & -T_4 & 0 & -T_7 & 0 & 0 & 0 & 0 & 0 \\ 0 & 0 & 0 & 0 & 0 & T_1 & 0 & -T_5 & 0 & -T_4 & 0 & 0 & 0 & 0 \\ 0 & 0 & 0 & 0 & 0 & 0 & 0 & 0 & 0 & 0 & 0 & 0 & 0 & 0 \\ 0 & 0 & 0 & 0 & 0 & 0 & 0 & 0 & 0 & 0 & 0 & 0 & 0 & 0 \end{bmatrix} \cdots.$$

Note that entries of $B(\phi)$ that result from interpretations of the nonlinear terms of $\phi$ become zero when passing to $\overline{B(\phi)}$. Thus, the nonzero columns of $\overline{B(\phi)}$ correspond precisely to the linear columns of $\phi$ and the 0-rows of $\overline{B(\phi)}$ correspond to vertices that do not appear as endpoints of any path of length two in $G$. Such vertices can be isolated, part of a connected component consisting of a single edge, or the center vertex of a connected component that is a tree of diameter 2. Deleting zero-rows and zero-columns will not change the minors of a matrix. Thus, in practice, when focusing on minors, one can work with a smaller matrix $\phi'$ defined by the linear columns of $\phi$, and assume that the content ideal, $I_1(\phi')$, is generated by a subset of the variables, say $x_1, \ldots, x_d$. In this case, $[T_1, \ldots, T_r] \cdot \phi' = [x_1, \ldots, x_d] \cdot \overline{B(\phi)}$.

When $I$ is a monomial ideal, a particular presentation matrix $\phi$ of $I$ that we make use of comes from the Taylor resolution of $I$. We now recall the construction of the Taylor resolution and its presentation matrix (see [33,34] for more details). For a collection $\mathcal{B} = \{f_1, \ldots, f_r\}$ of polynomials in $R$ and a subset $\sigma \subseteq \mathcal{B}$, let $f_\sigma$ denote the least common multiple of $\{f_i \mid f_i \in \sigma\}$.

**Definition 3.** *Let $I \subseteq R$ be a monomial ideal with the unique set of monomial generators $\mathcal{B} = \{f_1, \ldots, f_r\}$. The* Taylor resolution *of $I$ is the following complex:*

$$0 \to F_r \xrightarrow{\partial_r} F_{r-1} \xrightarrow{\partial_{r-1}} \cdots \xrightarrow{\partial_2} F_1 \xrightarrow{\partial_1} I \to 0,$$

*where, for $p = 1, \ldots, r$, $F_p = R^{\binom{r}{p}}$ is the free $R$-module of rank $\binom{r}{p}$ whose basis corresponds to all subsets of $p$ elements from $\mathcal{B}$, and the differential map $\partial_p : F_p \to F_{p-1}$ is defined, for each basis element $e_\sigma \in F_p$ corresponding to a subset $\sigma$ of cardinality $p$ in $\mathcal{B}$, by*

$$\partial_p(e_\sigma) = \sum_{f_\ell \in \sigma} (-1)^{|\{f_j \in \sigma \mid j < \ell\}|} \frac{f_\sigma}{f_{\sigma \setminus \{f_\ell\}}} e_{\sigma \setminus \{f_\ell\}}.$$

The presentation matrix *of $I$ from its Taylor resolution is the matrix corresponding to the map $F_2 \xrightarrow{\partial_2} F_1$; its $(\{f_j\}, \tau)$-entry, for $\{f_j\} \subseteq \mathcal{B}$, $\tau \subseteq \mathcal{B}$ with $|\tau| = 2$, is equal to 0 if $f_j \notin \tau$, equal to $(-1)\dfrac{f_\tau}{f_j}$ if $\tau = \{f_j, f_k\}$ and $j < k$, and equal to $\dfrac{f_\tau}{f_j}$ if $\tau = \{f_j, f_k\}$ and $j > k$.*

The matrix $\phi$ in Example 1 is an instance of the presentation matrix that comes from the Taylor resolution of a monomial ideal. Another important notion that we use is minors and ideals of minors of a matrix.

**Definition 4.** *Let $A$ be an $r \times s$ matrix whose entries are polynomials in $R$. For $t \leq \min\{r, s\}$, a $t \times t$ minor of $A$ is the determinant of a $t \times t$ submatrix of $A$. The ideal in $R$ generated by all $t \times t$ minors of $A$ is often denoted by $I_t(A)$. A minor is* binomial *if it can be written as the sum (or difference) of two monomials in $R$.*

**Graph theory.** An *undirected graph* $G = (V, E)$ consists of a set $V$ of distinct points, called the *vertices*, and a collection $E$ of unordered pairs of vertices, called the *edges*. We assume that all graphs in this paper are *simple*; that is, a graph has neither loops nor multiple edges. We write $xy$ for the undirected edge between vertices $x$ and $y$ in a graph.

**Definition 5.** *Let G be an undirected graph.*

1. *A walk is an alternating sequence of vertices and edges $x_1, e_1, x_2, e_2, \ldots, e_{s-1}, x_s$ such that $e_i = \{x_i, x_{i+1}\}$ for all $i = 1, \ldots, s-1$. Such a walk is said to be* closed *if $x_1 = x_s$.*
2. *A walk is called a* trail *if its edges are distinct (while its vertices may repeat). A closed trail is called a* circuit.
3. *A walk is called a* path *if its edges and vertices are distinct (except possibly at $x_1 = x_s$). A closed path is called a* cycle.
4. *The* length *of a walk is the number of edges that the walk transverses (including multiplicities). A walk is* even *(or* odd*) if its length is an even (or odd) number.*

We often list only the vertices to indicate a walk since the edges are obvious from the vertices. The main graph-theoretic structure that our work captures in this paper is indecomposable even circuits, which we define below. We also recall a similar notion of primitive even closed walks.

**Definition 6.** *Let G be a graph.*

1. *An even circuit is* indecomposable *if it cannot be realized as the edge-disjoint union of two smaller even circuits.*
2. *An even closed walk is* primitive *if it does not contain an even closed subwalk.*

**Remark 1.** *There is a close connection between even closed walks in a graph G and the equations of the Rees algebra of the edge ideal of G (see [35]). In particular, suppose $x_1, e_1, x_2, e_2, \ldots, e_{2s-1}, x_{2s}, e_{2s}$, where $e_i = \{x_i, x_{i+1}\}$ and $e_{2s} = \{x_{2s}, x_1\}$, is an even closed walk and $\theta(T_i) = e_i$ as in Definition 1. Then,*

$$\theta(\prod_{i=1}^{s} T_{2i} - \prod_{i=1}^{s} T_{2i-1}) = \prod_{i=1}^{s} e_{2i} - \prod_{i=1}^{s} e_{2i-1} = \prod_{i=1}^{2s} x_i - \prod_{i=1}^{2s} x_i = 0$$

*so $\prod_{i=1}^{s} T_{2i} - \prod_{i=1}^{s} T_{2i-1} \in J$.*

An application of our work is to directed graphs, thus we also recall basic terminology for directed graphs. A *digraph* $D = (Z, \vec{E})$ consists of a set $Z$ of distinct points, called the *vertices*, and a collection $\vec{E}$ of ordered pairs of vertices, called the *directed edges*. We also assume that all digraphs in this paper are simple digraphs. We write $x \to y$ for the directed edge from $x$ to $y$ in a digraph.

Directed walks, paths, circuits and cycles in a digraph can be defined similarly to those in an undirected graph with only one difference, that is, if $x_1, e_1, x_2, \ldots, e_{s-1}, x_s$ represents a directed walk from $x_1$ to $x_s$, then $e_i$ is the directed edge $x_i \to x_{i+1}$ for all $i = 1, \ldots, s-1$.

The application of our work to directed cycles in digraphs is based on the following construction [36].

**Definition 7.** *Let $D = (Z, \vec{E})$ be a digraph over the vertex set $Z = \{z_1, \ldots, z_m\}$. The bipartite graph $G(D)$, associated to D, is constructed as follows.*

1. *The vertices of $G(D)$ are $\{x_1, \ldots, x_m, y_1, \ldots, y_m\}$.*
2. *The edges of $G(D)$ are:*

    (a) *$\{x_i, y_i\}$ for all $i = 1, \ldots, m$; and*
    (b) *$\{x_i, y_j\}$, for $i \neq j$, if $z_i \to z_j$ is an edge in $\vec{E}$.*

It is easy to see that, for any digraph $D = (Z, \vec{E})$, the bipartite graph $G(D)$ has a perfect matching $\{e_i = \{x_i, y_i\} \mid i = 1, \ldots, m\}$. We denote this perfect matching of $G(D)$ by $M_D$.

**Example 2.** *Let D be directed graph*

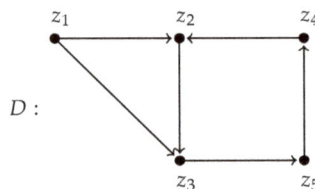

*Then, the bipartite graph $G = G(D)$ associated to D is*

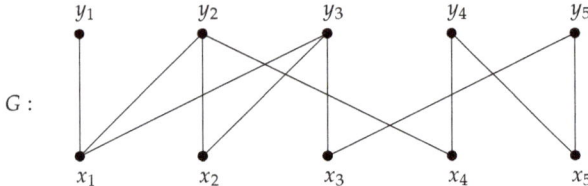

## 3. Even Circuits in Graphs

In this section, we present an algebraic algorithm to enumerate indecomposable even circuits in a graph. Recall that $G = (V, E)$ is a simple graph on the vertex set $V = \{x_1, \ldots, x_n\}$ with $r = |E|$. For $I = I(G)$, fix $\phi$ to be the presentation matrix of $I = I(G)$ that results from the Taylor resolution of $I$, as in Definition 3. For the remainder of the paper, $\phi$ always refers to the Taylor presentation matrix unless otherwise noted.

We start with the following simple observation about $\phi$. Example 1 already illustrates the statements below, which are generally known but written here for ease of reference.

**Lemma 1.** *If G is a graph and $I = I(G)$, then the following statements hold.*

1. *The entries of $\phi$ are monomials in $\{x_1, \ldots, x_n\}$.*
2. *Every column of $\phi$ has precisely two nonzero entries.*
3. *The nonzero entries in each column of $\phi$ are either both linear or both quadratic.*
4. *Every linear column of $\phi$ corresponds to a path of length 2 in G whose end-vertices are the nonzero entries of this column.*
5. *Every quadratic column of $\phi$ corresponds to a pair of disjoint edges.*

**Proof.** The assertions are straightforward from the construction of the Taylor resolution of $I(G)$. Note that, in general, all relations on a set of monomials can be generated by pairwise relations (i.e., relations of two monomials). If $m_1, m_2$ are monomials, then a minimal relation between them has the form $am_1 + bm_2 = 0$ where $a = (lcm(m_1, m_2)/m_1)$ and $b = (lcm(m_1, m_2)/m_2)$ are monomials. If $m_1 \neq m_2$ both have degree 2, then $a = m_2, b = m_1$ if $m_1, m_2$ have disjoint support. Otherwise, $a, b$ both have degree one. The results follow. □

As above, we denote by $B(\phi)$ and $\overline{B(\phi)}$ the Jacobian dual and the reduced Jacobian dual of $\phi$, respectively. We obtain an immediate corollary of Lemma 1 when $\phi$ is assumed to be the Taylor presentation matrix of $I(G)$ for a graph $G$.

**Corollary 1.** *The nonzero columns of $\overline{B(\phi)}$ are precisely the columns of $B(\phi)$ that correspond to linear columns of $\phi$. Particularly, each nonzero column of $\overline{B(\phi)}$ contains precisely two nonzero entries, each of which is a degree one monomial in $T_1, \ldots, T_r$.*

**Proof.** By definition, the nonzero entries of $\overline{B(\phi)}$ come from nonzero entries of $B(\phi)$ that are contained in $\Bbbk[T_1, \ldots, T_r]$. Observe, by the equation

$$[T_1 \ \ldots \ T_r] \cdot \phi = [x_1 \ \ldots \ x_n] \cdot B(\phi), \tag{1}$$

that the columns of $B(\phi)$ correspond to columns of $\phi$. Moreover, by Lemma 1, the nonzero entries in each column of $\phi$ are of the same degrees (either linear or quadratic). It further follows from Equation (1) that the degree with respect to the $x_i$s of nonzero entries of a column in $B(\phi)$ is exactly one less than that of the corresponding column of $\phi$. Hence, nonzero columns of $\overline{B(\phi)}$ correspond to columns without the $x_i$s in $B(\phi)$, which correspond to columns of linear forms (and 0) in $\phi$.

The second statement also follows from Lemma 1. □

**Remark 2.** *Note that, since zero columns of a matrix do not play any role in what follows, we could define $\overline{B(\phi)}$ to exclude all its zero columns. That is, we are working just with the (uniquely defined) columns of $\phi$, whose nonzero entries are linear, that result from binomial relations of edges in paths of length 2 in G. As mentioned above, zero rows of $\overline{B(\phi)}$ do not play a role and can be eliminated by using the content ideal of $\phi$ to define $B(\phi)$ rather than $\mathfrak{m}$. However, since zero rows do not affect minors, which are our main focus when using $\overline{B(\phi)}$, it is a matter of convenience to allow them.*

As stated in Lemma 1, the linear columns of $\phi$ are generated by pairs of monomials corresponding to edges that share a vertex. In other words, the linear columns of $\phi$ correspond to paths of length 2 in the graph. It can be desirable for computational purposes to use a minimal presentation matrix for $\phi$ rather than the full Taylor presentation matrix. It is easy to check that there are three paths of length two in each triangle, yielding three linear relations, any two of which generate the third. Since this is the only redundancy among the linear relations for a graph, if the graph is triangle free, the linear columns of a minimal presentation matrix will be the same as the linear columns of the Taylor presentation matrix.

Since the linear columns arise from paths of length two, as seen in Lemma 1, the endpoints of each path are the nonzero entries of that column of $\phi$. These endpoints will thus be encoded in the corresponding column of $B(\phi)$ as the rows in which the nonzero entries appear. It is natural to expect that the third vertex, the midpoint of the path, would play a role.

**Definition 8.** *We call two nonzero columns of $\overline{B(\phi)}$ center-distinct if their corresponding paths of length 2 in G have distinct middle vertices. We also call the middle vertices of these paths of length 2 the mid-points of the corresponding columns.*

Finding the mid-point of a column of $\overline{B(\phi)}$ can be done easily by examining the corresponding edges of $G$. If $T_i$ and $T_j$ are the two nonzero entries of a column of $\overline{B(\phi)}$ and $f_i, f_j$ are the corresponding edges of $G$ (that is, $\theta(T_i) = f_i t$, and $\theta(T_j) = f_j t$), then the mid-point of the column is $\operatorname{supp} f_i \cap \operatorname{supp} f_j$, or equivalently $\gcd(f_i, f_j)$.

The next lemma collects information that can be gleaned about a graph from minors of $\overline{B(\phi)}$ of a form that appears frequently in the remainder of the article.

**Lemma 2.** *If $\overline{B(\phi)}$ has a minor of the form*

$$M = \begin{bmatrix} T_2 & 0 & \cdots & -T_{2t-1} \\ -T_1 & T_4 & \cdots & 0 \\ 0 & -T_3 & \cdots & 0 \\ \vdots & \vdots & \cdots & \vdots \\ 0 & 0 & \cdots & T_{2t} \end{bmatrix}, \qquad (2)$$

*then $G$ contains an even closed walk corresponding to $\det M$. Moreover,*

1. *the walk is primitive if and only if the columns of $M$ are pairwise center-distinct; and*
2. *the walk is a circuit if and only if the nonzero entries of $M$ are distinct, in which case the circuit is indecomposable if and only if the columns of $M$ are pairwise center-distinct.*

**Proof.** Combining Equation (1) with Lemma 1 gives that each column of $M$ corresponds to a path of length 2 in $G$ whose end-vertices are labeled by the rows of $M$ corresponding to the nonzero entries in that column. By re-indexing the variables, we may assume that the rows of $M$ correspond to the variables $x_1, \ldots, x_t$. Then, the $i$th column of $M$, for $1 \leq i \leq t-1$, corresponds to a path of length 2 from $x_i$ to $x_{i+1}$, and the last column of $M$ corresponds to a path of length 2 from $x_t$ to $x_1$. We denote those paths by $x_i, y_i, x_{i+1}$, for $i = 1, \ldots, t-1$, and $x_t, y_t, x_1$. Furthermore, edges on these paths correspond to the variables $T_1, \ldots, T_{2t}$. Hence, these paths glue together to form an even closed walk of length $2t$ in $G$. Since $\det M = \prod_{i=1}^{t} T_{2i} - \prod_{i=1}^{t} T_{2i-1}$, we have that $\det M$ corresponds to an even closed walk in $G$, as in Remark 1. This walk is a circuit if and only if the $T_i$, and thus the corresponding edges, are distinct. Finally, the columns of $M$ are pairwise center-distinct if and only if the vertices $y_1, \ldots, y_t$ are pairwise distinct. Note that the vertices $x_1, \ldots, x_t$ are distinct by definition. This guarantees that the obtained closed walk or circuit of length $2t$ in $G$ is primitive or indecomposable, respectively, if and only if the columns of $M$ are pairwise center-distinct. □

We are now ready to prove the main result of the paper, Theorem 1. Note that relabeling the vertices or edges of a graph corresponds to rearranging the rows of $\phi$ or of $B(\phi)$. Such a rearrangement will not affect the minors of a matrix, thus, when convenient, a specific labeling of vertices can be used without loss of generality.

**Theorem 1.** *Let $G$ be a graph. Then, the indecomposable even circuits of $G$ correspond exactly to the binomial minors of $\overline{B(\phi)}$, which satisfy the following conditions:*

1. *the monomials in these binomials are square-free and relatively prime; and*
2. *the columns of the corresponding submatrices are pairwise center-distinct.*

**Proof.** Suppose that $C$ is an indecomposable even circuit in $G$. For ease of notation, select a labeling on the vertices and edges so that the edges of $C$ (in order) are $e_1, \ldots, e_{2t}$, where $e_i = x_i x_{i+1}$ for $i < 2t$ and $e_{2t} = x_{2t} x_1$. Since $C$ is indecomposable, it is easy to see that $x_1, x_3, \ldots, x_{2t-1}$ are pairwise distinct. Particularly, the linear relations of $I = I(G)$ include $x_1 e_2 - x_3 e_1, x_3 e_4 - x_5 e_3, \ldots, x_{2t-1} e_{2t} - x_1 e_{2t-1}$ which correspond to the following columns of $\phi$:

$$\begin{bmatrix} -x_3 & 0 & \cdots & 0 \\ x_1 & 0 & \cdots & 0 \\ 0 & -x_5 & \cdots & 0 \\ 0 & x_3 & \cdots & 0 \\ \vdots & \vdots & \cdots & \vdots \\ \vdots & \vdots & \cdots & -x_1 \\ 0 & 0 & \cdots & x_{2t-1} \\ \vdots & \vdots & \cdots & \vdots \end{bmatrix}$$

where, for convenience, labelings were chosen so that $T_i$ corresponds to $e_i$ for $1 \leq i \leq 2t$.

We can reorder the columns of $\phi$ so that these are the first $t$ columns. These columns produce $x_1 T_2 - x_3 T_1, x_3 T_4 - x_5 T_3, \ldots, x_{2t-1} T_{2t} - x_1 T_{2t-1}$ as linear equations of the Rees algebra $R[It]$, which correspond to the first $t$ equations of $[x_1 \ldots x_n] \cdot B(\phi)$. Thus, the first $t$ columns of $B(\phi)$ are:

$$\begin{bmatrix} T_2 & 0 & \ldots & -T_{2t-1} \\ 0 & 0 & \ldots & 0 \\ -T_1 & T_4 & \ldots & 0 \\ 0 & 0 & \ldots & 0 \\ 0 & -T_3 & \ldots & 0 \\ \vdots & \vdots & \ldots & \vdots \\ 0 & 0 & \ldots & T_{2t} \\ \vdots & \vdots & \ldots & \vdots \end{bmatrix}.$$

By Corollary 1, these columns of $B(\phi)$ are unchanged when passing to $\overline{B(\phi)}$. Consider the $t \times t$ submatrix $M$ of $\overline{B(\phi)}$ consisting of the first $t$ columns and the $t$ identified nonzero rows:

$$M = \begin{bmatrix} T_2 & 0 & \ldots & -T_{2t-1} \\ -T_1 & T_4 & \ldots & 0 \\ 0 & -T_3 & \ldots & 0 \\ \vdots & \vdots & \ldots & \vdots \\ 0 & 0 & \ldots & T_{2t} \end{bmatrix}.$$

Then, $\det(M) = \prod_{i=1}^{t} T_{2i} - \prod_{i=1}^{t} T_{2i-1}$. This is a binomial whose monomials are square-free and relatively prime. Observe further that, since $C$ is indecomposable, $x_2, x_4, \ldots, x_{2t}$ are pairwise distinct. Therefore, the columns of $M$ are pairwise center-distinct.

Conversely, suppose that $M$ is a $t \times t$ submatrix of $\overline{B(\phi)}$ whose determinant is a binomial of degree $t$ with square-free and relatively prime monomials, and whose columns are pairwise center-distinct. It follows from Lemma 1 and Corollary 1 that each column of $M$ contains at most two nonzero entries. Since the monomials in $\det(M)$ are relatively prime, each column of $M$ must contain exactly two nonzero entries. Particularly, $M$ contains exactly $2t$ nonzero entries. In addition, since the monomials in $\det(M)$ are relatively prime, $\det(M)$ contains exactly $2t$ distinct variables. Thus, all the $2t$ nonzero entries of $M$ are distinct. Since each row also contains at least two distinct entries in order for the monomials to be relatively prime, a simple counting argument guarantees exactly two nonzero entries per row as well.

Now, by rearranging the rows and columns of $M$, it is easy to put $M$ in a block-matrix form, where each nonzero block is of the form in Equation (2) and lies on the diagonal. Observe further that, if $M$ has more than one such block, then $\det(M)$ is not a binomial since all entries of $M$ are distinct. Therefore, we can assume that $M$ is exactly as in Equation (2). The result now follows from Lemma 2. □

Theorem 1 gives us the following algebraic algorithm to detect the existence of and list all even circuits in a given graph $G$.

**Algorithm 1** To enumerate all indecomposable even circuits in a given graph G

1. Form $\phi$.
2. Compute $\overline{B(\phi)}$.
3. for $t$ from 1 to the rank of $\overline{B(\phi)}$ compute all $t \times t$ minors of $\overline{B(\phi)}$.
4. Test if each minor satisfies the following conditions:
   (a) its columns are pairwise center-distinct; and
   (b) its determinant is a binomial whose monomials are square-free and relatively prime.
5. If the answer is "Yes", then return the rows and centers of the columns corresponding to the minor. These are the vertices of an indecomposable even circuit in G.

**Remark 3.** *Note that only the linear columns of $\phi$ are necessary in this process and so in Step 1 only the linear relations need be considered. Note also that the Taylor resolution and its presentation matrix can be constructed in polynomial time on the number of generators of $I(G)$ (i.e., the number of edges in G). Note further that the computation of the determinant of a matrix can also be done in polynomial time on the size of the matrix, and the rank of $\overline{B(\phi)}$ is at most the number of edges in G. Finally, testing if the columns of a minor in $\overline{B(\phi)}$ are center-distinct can be done in polynomial time on the size of the minor, which is bounded by the number of edges in G. Thus, Algorithm 1 runs in polynomial time on the size of the edge set of G.*

**Example 3.** *Consider the following graph.*

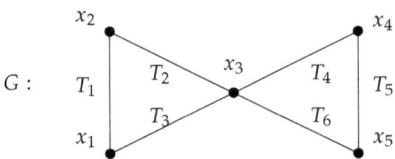

*The nonzero columns of the reduced Jacobian dual of G corresponding to the Taylor presentation matrix $\phi$ are computed to be:*

$$\overline{B(\phi)}: \begin{array}{c} x_1 \\ x_2 \\ x_3 \\ x_4 \\ x_5 \end{array} \left[ \begin{array}{cccccccccc} -T_2 & T_2 & 0 & 0 & 0 & 0 & 0 & 0 & -T_4 & -T_6 \\ 0 & -T_3 & T_3 & 0 & 0 & 0 & -T_4 & -T_6 & 0 & 0 \\ T_1 & 0 & -T_1 & -T_5 & T_5 & 0 & 0 & 0 & 0 & 0 \\ 0 & 0 & 0 & 0 & -T_6 & T_6 & T_2 & 0 & T_3 & 0 \\ 0 & 0 & 0 & T_4 & 0 & -T_4 & 0 & T_2 & 0 & T_3 \end{array} \right],$$

*where the labels $x_1, \ldots, x_5$ indicate the variables of the corresponding rows. Furthermore, the mid-points of the columns are successively $x_2, x_3, x_1, x_4, x_5, x_3, x_3, x_3, x_3, x_3$. By evaluating the minors of $\overline{B(\phi)}$, the only binomial minor whose monomials are square-free and relatively prime is $T_1T_4T_6 - T_2T_3T_5$, which corresponds to the only indecomposable even circuit $x_1, x_2, x_3, x_4, x_5, x_3, x_1$ in G. This minor appears using the submatrix formed by taking Rows 1, 3, 4 and Columns 1, 5, 9, for example, or the one formed by Rows 2, 3, 4 and Columns 3, 5, 7.*

**Remark 4.** *There can be binomial minors of $\overline{B(\phi)}$ whose monomials are neither square-free nor relatively prime. These minors may correspond to even closed walks which transverse an edge multiple times.*

**Example 4.** *Consider the following graph.*

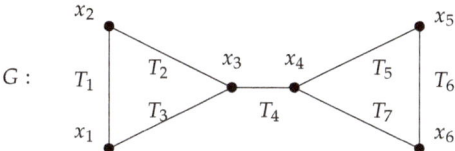

The nonzero columns of the reduced Jacobian dual of $I(G)$ with respect to the Taylor presentation matrix $\phi$ are computed to be:

$$\overline{B(\phi)}: \begin{array}{c} x_1 \\ x_2 \\ x_3 \\ x_4 \\ x_5 \\ x_6 \end{array} \left[ \begin{array}{cccccccccc} -T_2 & T_2 & 0 & 0 & -T_4 & 0 & 0 & 0 & 0 & 0 \\ 0 & -T_3 & T_3 & T_4 & 0 & 0 & 0 & 0 & 0 & 0 \\ T_1 & 0 & -T_1 & 0 & 0 & -T_5 & -T_7 & 0 & 0 & 0 \\ 0 & 0 & 0 & -T_2 & T_3 & 0 & 0 & -T_6 & T_6 & 0 \\ 0 & 0 & 0 & 0 & 0 & T_4 & 0 & 0 & -T_7 & T_7 \\ 0 & 0 & 0 & 0 & 0 & 0 & T_4 & T_5 & 0 & -T_5 \end{array} \right],$$

where the labels $x_1, \ldots, x_6$ indicate the variables of the corresponding rows. Furthermore, the mid-points of the columns are successively $x_2, x_3, x_1, x_3, x_3, x_4, x_4, x_5, x_6, x_4$.

The only binomial minor of $\overline{B(\phi)}$ is $T_2T_3T_5T_7 - T_1T_4^2T_6$. This corresponds to submatrices formed using Rows 1, 3, 4, 5 and Columns 1, 5, 6, 9 for example, or Rows 2, 3, 4, 6 and Columns 3, 4, 7, 8. A monomial of this minor is not square-free. This indicates that $G$ contains an even closed walk $x_2, x_3, x_4, x_5, x_6, x_4, x_3, x_1, x_2$, but this walk is not a circuit because it transverses through the edge $T_4$ twice. Hence, $G$ has no indecomposable even circuits.

**Remark 5.** With basically the same proof as that of Theorem 1, it can be shown that the even cycles of $G$ correspond exactly to the binomial minors of $\overline{B(\phi)}$, which satisfy the following conditions:

1. the monomials in these binomials are square-free and relatively prime; and
2. the variables labeling the rows and the mid-points of the columns of the corresponding submatrices are pairwise distinct.

**Example 5.** Let $G$ be the graph in Example 3. As shown in Example 3, the only binomial minor of $\overline{B(\phi)}$ whose monomials are square-free and relatively prime is $T_1T_4T_6 - T_2T_3T_5$. This minor is obtained by taking Rows 1, 3, 4 and Columns 1, 5, 9. In this minor, the mid-points of the columns are $x_2, x_5$ and $x_3$. On the other hand, the variables labeling the rows are $x_1, x_3$ and $x_4$. Clearly, we have a repeated $x_3$ among the mid-points and the row labels. Thus, this minor corresponds to an even indecomposable circuit, which is not a cycle.

**Example 6.** Let $G$ be the graph in Example 1. A binomial minor of $\overline{B(\phi)}$ whose monomials are square-free and relatively prime is $T_2T_5T_7 - T_1T_3T_6$. This minor is obtained by taking Rows 1, 3, 5 and Columns 1, 5, 9. The mid-points of the columns are $x_2, x_4$ and $x_6$, and the variables labeling the rows are $x_1, x_3$ and $x_5$. Since these are distinct variables, this minor corresponds to an even cycle $x_1, x_2, x_3, x_4, x_5, x_6, x_1$.

We continue this section by discussing an algebraic consequence of Theorem 1 in finding defining equations for the Rees algebras of edge ideals of graphs. Recall that the Rees algebra $R[It]$ of $I$ has a presentation $R[T_1, \ldots, T_r] \xrightarrow{\theta} R[It] \to 0$, and $J = \ker \theta$ is called the defining ideal of $R[It]$.

It was shown in [35] that the nonlinear equations of $J$ arise from the even closed walks in the graph $G$. An alternate proof of this fact appears in Chapter 10.1 of [24]. In addition, it was proved in [Corollary 10.1.5] of [24] that the generators correspond to primitive even closed walks and form a reduced Gröbner basis for $J$. The binomials corresponding to indecomposable even circuits of $G$ are thus known to be elements of $J$. However, there are elements of $J$ that do not correspond to indecomposable circuits, as seen in Example 4. It is worth noting that it was established in a more general setting that the maximal minors of $\overline{B(\phi)}$ are contained in $J$ (see, for example, [31]). A close examination of the proof of Theorem 1 shows that any even closed walk in $G$ corresponds to a binomial minor of $\overline{B(\phi)}$.

**Corollary 2.** *Let $G$ be a graph with edge ideal $I = I(G)$, and let $J$ be the defining ideal of $R[It]$. Then, every nonlinear generator of $J$ appears as a binomial minor of $\overline{B(\phi)}$.*

**Proof.** By the work in [35] and (Chapter 10.1 [24]), we have that the nonlinear generators of $J$ correspond to primitive even closed walks in $G$. Consider any primitive even closed walk $W$ in $G$ and, after a re-labeling, suppose that the vertices on $W$ are $x_1, \ldots, x_{2t}$ (not necessarily distinct).

Observe that, since $W$ is primitive, if $x_i = x_j$ is a repeated vertex in $W$, then $i$ and $j$ are of different parity. View $W$ as the union of $t$ paths of length 2, namely, $P_i = x_{2i-1}, x_{2i}, x_{2i+1}$, for $i = 1, \ldots t$ (where $x_{2t+1} = x_1$). Then, the endpoints of each path $P_i$ are distinct vertices. Thus, $P_i$ corresponds to a column of $\overline{B(\phi)}$ with exactly two nonzero entries, appearing in the rows labeled by $x_{2i-1} \neq x_{2i+1}$. Selecting these columns and the corresponding nonzero rows results in a $t \times t$ submatrix $M_W$ of $\overline{B(\phi)}$.

As in the proof of Theorem 1, the rows and columns of $M_W$ can be rearranged so that $M_W$ is a block-matrix in which each block is of the form of Equation (2). If there are multiple blocks, then each corresponds to an even closed walk contained in $W$ by Lemma 2, a contradiction to the fact that $W$ is primitive. Therefore, $M_W$ is of the form in Equation (2), where the nonzero entries may not be distinct. Hence, the corresponding generator of $J$ is the same as $\det(M_W)$, which is a binomial minor of $\overline{B(\phi)}$. □

Let $\mathbf{T} = [T_1 \ \ldots \ T_r]$. Corollary 2 gives us the containment

$$J \subseteq \langle \mathbf{T} \cdot \phi, I_2(\overline{B(\phi)}), I_3(\overline{B(\phi)}), \ldots, I_k(\overline{B(\phi)}) \rangle,$$

where $k$ is the rank of $\overline{B(\phi)}$. The reverse containment fails to hold. In general, $I_t(\overline{B(\phi)})$ will contain monomials that are not in $J$. For instance, in Example 1, $T_2 T_3 \in I_2(\overline{B(\phi)})$ but $T_2 T_3 \notin J$. Interestingly, we see that by restricting to binomial minors we in fact obtain an equality. While not all binomial minors of $\overline{B(\phi)}$ are minimal generators of $J$, such minors correspond to multiples of binomials which come from (not necessarily primitive) even closed walks and are elements of $J$. The following lemma is used in proving the desired equality. For convenience, we consider 1 to be a trivial monomial.

**Lemma 3.** *Suppose $\psi$ is an $n \times n$ matrix such that:*

1. *$\det(\psi)$ is a nonzero binomial; and*
2. *every column of $\psi$ has at most 2 nonzero entries.*

*Then, after row and column rearrangements, $\psi$ has a block decomposition*

$$\psi = \begin{bmatrix} X & W \\ Y & Z \end{bmatrix}$$

*where $\det(X)$ is a monomial, $\det(Z)$ is a binomial, $\det(\psi) = \det(X)\det(Z)$, and every row of $Z$ has at least two nonzero entries and every column of $Z$ has exactly two nonzero entries.*

**Proof.** If every row of $\psi$ has at least two nonzero entries and every column of $\psi$ has exactly two nonzero entries, set $X, Y, W$ to be empty matrices and $Z = \psi$. Then, since an empty product is defined to be 1, $\det(X) = 1$ is a (degenerate) monomial, and the result holds. In particular, the result holds when $n = 2$. Assume $n > 2$.

Assume there exist $s$ rows of $\psi$ with a single nonzero entry. Since $\det(\psi) \neq 0$, every row and column of $\psi$ has at least one nonzero entry and no two rows (columns) have a single nonzero entry in the same column (row). Note also that row and column exchanges modify only the sign of the determinant and not the binomial nature. By performing row exchanges, we can rearrange all rows with a single nonzero entry to come before all rows with multiple nonzero entries. That is, we may assume that $\psi$ has the form

$$\psi = \left[\begin{array}{c|c} D_1 & 0 \\ \hline A_1 & \psi_1 \end{array}\right],$$

where $D_1$ is an $s \times s$ diagonal matrix, and $A_1$ is a matrix where each column has at most one nonzero entry. Observe that $\det(D_1)$ is a monomial and $\det(\psi) = \det(D_1)\det(\psi_1)$, thus $\det(\psi_1)$ is again a nonzero binomial. As before, each column of $\psi_1$ has either one or two nonzero entries and each row has at least one nonzero entry. Since $\psi_1$ is $n - s \times n - s$ with $s \geq 1$, by induction,

$$\psi_1 = \left[\begin{array}{cc} X_1 & W_1 \\ Y_1 & Z_1 \end{array}\right]$$

where $\det(X_1)$ is a monomial, $\det(Z_1)$ is a binomial, $\det(\psi_1) = \det(X_1)\det(Z_1)$, and every row of $Z_1$ has at least two nonzero entries and every column of $Z_1$ has exactly two nonzero entries. Now,

$$\psi = \left[\begin{array}{c|cc} D_1 & \multicolumn{2}{c}{0} \\ \hline A_1 & X_1 & W_1 \\ & Y_1 & Z_1 \end{array}\right] = \left[\begin{array}{c|c|c} D_1 & 0 & 0 \\ \hline A_1' & X_1 & W_1 \\ \hline A_1'' & Y_1 & Z_1 \end{array}\right],$$

where $A_1', A_1''$ consist of the appropriate rows of $A_1$. Set $Z = Z_1$, $X = \left[\begin{array}{c|c} D_1 & 0 \\ \hline A_1' & X_1 \end{array}\right]$, $W = \left[\begin{array}{c} 0 \\ \hline W_1 \end{array}\right]$, and $Y = \left[\begin{array}{c|c} A_1'' & Y_1 \end{array}\right]$. Note that

$$\det(\psi) = \det(D_1)\det(\psi_1) = \det(D_1)\det(X_1)\det(Z_1) = \det(X)\det(Z)$$

and the result follows.

Similarly, if any column of $\psi$ has a single nonzero entry, then by performing column exchanges, we may assume $\psi$ has the form

$$\psi = \left[\begin{array}{c|c} D_2 & B_2 \\ \hline 0 & \psi_2 \end{array}\right],$$

where $D_2$ is a diagonal matrix. Observe that $\det(\psi) = \det(D_1)\det(\psi_2)$ and $\det(D_2)$ is a monomial, thus $\det(\psi_2)$ is again a nonzero binomial. As before, each column of $\psi_2$ has either one or two nonzero entries and each row has at least one nonzero entry and the result follows by induction as in the case above. □

Note that in the above lemma, since the columns of $Z$ have two nonzero entries each and the columns of $\psi$ have at most two nonzero entries, it follows that $W = 0$. To state our next result formally, for a matrix $M$, set $\mathrm{bi}(I_t(M))$ to be the collection of $t \times t$ minors of $M$ that are binomials.

**Theorem 3.** *Let $G$ be a graph with edge ideal $I = I(G)$. Let $J$ be the defining ideal of $R[It]$ and let $k = \mathrm{rank}\,\overline{B(\phi)}$. Then,*

$$J = \langle \mathbf{T} \cdot \phi, \mathrm{bi}(I_2(\overline{B(\phi)})), \mathrm{bi}(I_3(\overline{B(\phi)})), \ldots, \mathrm{bi}(I_k(\overline{B(\phi)})) \rangle.$$

**Proof.** One inclusion follows directly from Corollary 2.

For the reverse inclusion, suppose that $\psi$ is a $t \times t$ submatrix of $\overline{B(\phi)}$ with $\det(\psi)$ a binomial. We show that $\det(\psi) \in J$. Indeed, since $\det(\psi) \neq 0$, every row and column of $\psi$ has at least one nonzero entry and no two rows (columns) have a single nonzero entry in the same column (row). As noted before, each nonzero column of $\overline{B(\phi)}$ has precisely two nonzero entries. Thus, each column of $\psi$ has at most two nonzero entries. Applying Lemma 3, we can assume that $\psi = \left[\begin{array}{c|c} A & B \\ \hline C & \psi_p \end{array}\right]$ where $\psi_p$ is a minor of $\psi$ in which every column has exactly two nonzero entries, every row has at least two nonzero entries, and $\det(\psi_p)$ is a nonzero binomial with $\det(\psi) = \det(A)\det(\psi_p)$. Thus, if $\det(\psi_p) \in J$, then $\det(\psi) \in J$.

Now, we can reorder the rows of $\psi_p$ so that the nonzero entries of the first column appear in the first two rows. Since the second row has at least two nonzero entries, we can rearrange the remaining columns of $\psi_p$ so that the $(2,2)$ entry is not zero. If the second nonzero entry of Column 2 is not in Row 1, then we can rearrange the remaining rows so that the $(3,2)$ entry of $\psi_p$ is not zero. We can continue to rearrange the rows and columns of $\psi_p$ in this manner (see also Equation (2)) until for some row $i$, the remaining columns with nonzero entries in row $i$ have the second nonzero entry in row $j$ for some $j < i$. At this point, $\psi_p$ has the following form

$$\psi_p = \left[\begin{array}{cccccc|c} * & 0 & 0 & 0 & \cdots & 0 & \\ * & * & 0 & 0 & \cdots & 0 & \\ 0 & * & * & 0 & \cdots & * & \\ 0 & 0 & * & * & \cdots & 0 & N \\ & & & \ddots & & & \\ 0 & 0 & 0 & 0 & \cdots & * & \\ \hline & & & 0 & & & \psi_{p+1} \end{array}\right],$$

where $*$ denotes a nonzero entry and the position of the second $*$ in the final column before $N$ is illustrative. Notice that $\psi_p$ has a minor of the form of Equation (2). In addition, $\psi_p$ has the form

$$\psi_p = \left[\begin{array}{c|c|c} L_{\psi_p} & 0 & \\ \hline C_{\psi_p} & M_{\psi_p} & N \\ \hline & 0 & \psi_{p+1} \end{array}\right],$$

where $L_{\psi_p}$ is lower triangular (the empty matrix if $j = 1$), and $M_{\psi_p}$ has the form of Equation (2). Note that

$$\det(\psi_p) = \det(L_{\psi_p})\det(M_{\psi_p})\det(\psi_{p+1}).$$

By Lemma 2, $\det(M_{\psi_p}) \in J$, and since $J$ is closed under multiplication, $\det(\psi) \in J$ as required. □

**Remark 6.** *It can be seen in the proof above that every row of $\psi_{p+1}$ has at least two nonzero entries and each column has either one or two nonzero entries. We can continue the process to get a block decomposition of $\psi$, in which the determinant is the product of the determinants of the diagonal blocks, each of which is either a monomial or a binomial coming from a block of the form of Equation (2). Hence, $\det(\psi)$ can be written as a product of monomials and of binomials coming from blocks of the form of Equation (2). Since minors of the form of Equation (2) correspond to (not necessarily primitive) even closed walks, these binomials are in $J$ by [35]. Thus, not only are the binomial minors in $J$, if they are not irreducible, they factor as products of monomials and binomial elements of $J$.*

In the proof of Theorem 3, the fact that $\det(\psi)$ was a binomial came from the statement. Since the entries of $\psi$ are not assumed to be distinct, it is possible for the product of two binomials to be a binomial. However, much of the proof holds if instead $\det(\psi)$ is assumed not to be a monomial. The following result demonstrates how to use this to obtain all nonlinear generators of $J$ from a single sized ideal of minors of $\overline{B(\phi)}$.

**Corollary 3.** *Let $k = \text{rank } \overline{B(\phi)}$. Then, all nonlinear generators of $J$ can be obtained as factors of generators of $I_k(\overline{B(\phi)})$.*

**Proof.** By Corollary 2, every nonlinear generator of $J$ appears as some binomial minor of $\overline{B(\phi)}$. Let $f$ be a nonlinear generator of $J$ and let $M$ be the corresponding submatrix of $\overline{B(\phi)}$. Since $J$ is generated by primitive even walks, the monomials in $f$ are relatively prime, thus every column of $M$ contains exactly two nonzero entries. By performing row and column exchanges, write

$$\overline{B(\phi)} = \left[\begin{array}{c|c} M & * \\ \hline 0 & B_2 \end{array}\right].$$

The set of columns of $\overline{B(\phi)}$ used to form $M$ can be extended to a set of columns of full rank. That is, by selecting appropriate columns and rows from $B_2$, there is a $k \times k$ submatrix of $\overline{B(\phi)}$ of the form

$$\widehat{M} = \left[\begin{array}{c|c} M & * \\ \hline 0 & B_2' \end{array}\right]$$

with nonzero determinant. Since $\det(\widehat{M}) = \det(M)\det(B_2')$, $f$ is a factor of an element of $I_k(\overline{B(\phi)})$ as desired. □

## 4. Directed Cycles in Digraphs

In this part of the paper, we discuss an interesting application of our main result, Theorem 1, to the problem of detecting the existence of directed cycles in a given directed graph.

Let $D = (Z, \vec{E})$ be a directed graph over the vertex set $Z = \{z_1, \ldots, z_m\}$. Let $G = G(D)$ be the undirected bipartite graph constructed from $D$ as in Definition 7. Recall that $M_D$ represents the perfect matching $\{e_i = x_i y_i \mid i = 1, \ldots, m\}$ in $G = G(D)$.

The connection between directed cycles in a digraph $D$ and even cycles in $G(D)$ is well established (see, for example, [22,23]). We present the following known result in a form that is convenient for applying Theorem 1. Note that, in an even circuit, we can start at any place and collect the first, third, fifth, etc. (all the odd ordered) edges, or collect the second, fourth, etc. (all the even ordered) edges. This way we get two sets of disjoint edges, each consisting of exactly half the number of edges on the circuit. We refer to each of these two sets a *collection of alternating edges* in the even circuit.

**Theorem 2.** The directed cycles in $D$ correspond exactly to the even cycles in $G = G(D)$ in which a collection of alternating edges forms a subset of the perfect matching $M_D$ in $G$.

**Proof.** We start the proof with the following observation. Consider an even circuit $C$ in $G$ with the property that its alternating edges form a subset of the perfect matching $M_D$. It can be seen that, if one transverses around $C$ on its edges and hits $x_i$ (or $y_i$) from an edge that is not in $M_D$, then the next edge of $C$ has to be $x_i y_i$. Since in a circuit the edges are distinct, this ensures that $C$ cannot contain $x_i$ (or $y_i$) more than once. That is, $C$ is a cycle (which is necessarily indecomposable). Thus, the indecomposable circuits in $G$ with the property that a collection of their alternating edges forms a subset of the perfect matching $M_D$ are exactly the even cycles in $G$ with the same property.

Suppose that $z_{i_1} \to z_{i_2} \to \cdots \to z_{i_t} \to z_{i_1}$ is a directed cycle in $D$. By the construction of $G$, it is easy to see that $x_{i_1}, y_{i_2}, x_{i_2}, y_{i_3}, x_{i_3}, \ldots, x_{i_t}, y_{i_1}, x_{i_1}$ is an indecomposable even circuit in $G$. Moreover, the collection of even edges in this circuit is $\{e_{i_1}, \ldots, e_{i_t}\}$, which is a subset of the perfect matching $M_D$.

Conversely, suppose that $G$ contains an indecomposable even circuit $C$ in which a collection of alternating edges form a subset of the perfect matching $M_D$. Since $G$ is bipartite, every edge in $G$ (and so any edge in $C$) connects a vertex $x_{i_j}$ to a vertex $y_{i_k}$. Thus, without loss of generality, we may assume that the circuit $C$ is of the form $x_{i_1}, y_{i_2}, x_{i_2}, y_{i_3}, x_{i_3}, \ldots, x_{i_t}, y_{i_1}, x_{i_1}$. Since $C$ is a circuit and $C$ contains $\{e_{i_1}, \ldots, e_{i_t}\}$, it follows that $i_1, \ldots, i_t$ are distinct indices. By the construction of $G$ again, we have a directed cycle $z_{i_1} \to z_{i_2} \to \cdots \to z_{i_t} \to z_{i_1}$ in $D$. □

**Example 7.** *Let $D$ be the directed graph in Example 2 and let $G = G(D)$ be its associated bipartite graph. It can be seen that $G$ has only one even cycle whose alternating edges form a subset of the perfect matching $M_D$, namely,*

$x_2, y_3, x_3, y_5, x_5, y_4, x_4, y_2, x_2$. This even cycle of G corresponds to the directed cycle $z_2 \to z_3 \to z_5 \to z_4 \to z_2$ in D.

Note that $z_1, z_2, z_3, z_1$ does not form a directed cycle in D even though its undirected edges would form a triangle. This is reflected by the fact that there is no even cycle between $x_1, y_1, x_2, y_2, x_3, y_3$ in G. Furthermore, not all even cycles in G would correspond to directed cycles in D. For instance, consider the even cycle $x_1, y_3, x_3, y_5, x_5, y_4, x_4, y_2, x_1$ in G. Neither collection of alternating edges of this cycle is a subset of the perfect matching $M_D$, and this even cycle does not correspond to any directed cycle in D ($z_1, z_3, z_5, z_4, z_2, z_1$ does not form a directed cycle in D).

As a corollary of Theorem 2, we derive an algebraic algorithm to enumerate all directed cycles in a given digraph. Note that, by the proof of Theorem 2, indecomposable even circuits of $G = G(D)$, in which a collection of alternating edges is a subset of $M_D$, are exactly the even cycles in G with the same property.

**Corollary 4.** *Let D be a digraph and let $G = G(D)$ be its corresponding bipartite graph. Let $\phi$ be the presentation of the edge ideal $I = I(G)$ of G that results from its Taylor resolution. Let $\overline{B(\phi)}$ be its reduced Jacobian dual. Then, the directed cycles of length t in D correspond exactly to the binomial $t \times t$ minors of $\overline{B(\phi)}$ that satisfy the following conditions:*

1. *their columns are pairwise center-distinct;*
2. *their monomials are square-free and relatively prime; and*
3. *one of these monomials is the product of variables that correspond to a subset of the perfect matching $M_D$.*

**Proof.** The assertion is a direct consequence of Theorems 1 and 2. □

We recall the famous Caccetta–Häggkvist conjecture for directed cycles in digraphs [37].

**Conjecture 4** (Caccetta–Häggkvist). *Let D be a digraph on n vertices. Let $\ell \in \mathbb{N}$ and suppose that the outdegree of each vertex in D is at least $\frac{n}{\ell}$. Then, D contains a directed cycle of length at most $\ell$.*

As a consequence of Theorems 1 and 2, we are able to present a Jacobian dual matrix interpretation of the Caccetta–Häggkvist conjecture as follows. Note that every cycle is an indecomposable even circuit, and that, for a bipartite graph, the two notions coincide.

**Conjecture 5.** *Let D be a digraph on n vertices such that the outdegree of each vertex in D is at least $\frac{n}{\ell}$. Let $\phi$ be the Taylor presentation matrix of $I(G(D))$. Then, for some $q \leq \ell$, $I_q(\overline{B(\phi)})$ contains a binomial with square-free, relatively prime terms, one of which is a product of elements of $M_D$.*

By Theorems 1 and 2, Conjectures 4 and 5 are equivalent. Conjecture 5 can also be rephrased using the language of Rees algebras by using Theorem 3.

**Conjecture 6.** *Let D be a digraph on n vertices such that the outdegree of each vertex in D is at least $\frac{n}{\ell}$. If J is the defining ideal of the Rees algebra $R[I(G(D))t]$, then, for some $q \leq \ell$, J has a binomial generator of degree q that is square-free and has relatively prime terms, one of which is a product of elements of $M_D$.*

We conclude the paper with the observation that Conjecture 5 can be further translated into a problem in linear algebra. Notice that, if the outdegree of a vertex $z_i$ is at least r, then there are at least r paths of length 2 using the edge $x_i y_i$ with $x_i$ as the mid-point. The corresponding linear relations $y_i T_{j_i} - y_j T_i$ give specific information about r columns of $\overline{B(\phi)}$, each of which has an element from the perfect matching. If D has m vertices, this yields mr columns of $\overline{B(\phi)}$, each of which contains an element from the perfect matching, which form a fertile source of potential minors using submatrices of the form of Equation (2) that would correspond to directed cycles in D.

**Author Contributions:** Conceptualization, H.T.H. and S.M.; Formal analysis, H.T.H. and S.M.; Investigation, H.T.H. and S.M.; Methodology, H.T.H. and S.M.; Writing–original draft, H.T.H. and S.M.; Writing–review and editing, H.T.H. and S.M. Both authors contributed equally to this work.

**Funding:** The first author is partially supported by Louisiana Board of Regents (grant #LEQSF(2017-19)-ENH-TR-25).

**Acknowledgments:** The authors thank the anonymous referees for a careful, detailed reading of the manuscript and useful suggestions.

**Conflicts of Interest:** The authors declare no conflict of interest.

## References

1. Alon, N.; Yuster, R.; Zwick, U. *Finding and Counting Given Length Cycles (Extended Abstract)*; Algorithms-ESA '94 (Utrecht), Lecture Notes in Comput. Sci. 855; Springer: Berlin, Germany, 1994; pp. 354–364,.
2. Alon, N.; Yuster, R.; Zwick, U. Finding and counting given length cycles. *Algorithmica* **1997**, *17*, 209–223. [CrossRef]
3. Chudnovsky, M.; Kawarabayashi, K.-I.; Seymour, P. Detecting even holes. *J. Graph Theory* **2005**, *48*, 85–111. [CrossRef]
4. Conforti, M.; Cornuéjols, G.; Kapoor, A.; Vušković, K. Even-hole-free graphs. II. Recognition algorithm. *J. Graph Theory* **2002**, *40*, 238–266. [CrossRef]
5. Conlon, J.G. Even cycles in graphs. *J. Graph Theory* **2004**, *45*, 163–223. [CrossRef]
6. Dahlgaard, S.; Knudsen, M.B.T.; Stöckel, M. Finding even cycles faster via capped K-walks. In *STOC'17—Proceedings of the 49th Annual ACM SIGACT Symposium on Theory of Computing*; ACM: New York, NY, USA, 2017; pp. 112–120.
7. Fan, G.; Zhang, C.-Q. Circuit decompositions of Eulerian graphs. *J. Comb. Theory Ser. B* **2000**, *78*, 1–23. [CrossRef]
8. Huynh, T.; Oum, S.; Verdian-Rizi, M. Even-cycle decompositions of graphs with no odd-$K_4$-minor. *Eur. J. Comb.* **2017**, *65*, 1–14. [CrossRef]
9. Huynh, T.; King, A.D.; Oum, S.; Verdian-Rizi, M. Strongly even-cycle decomposable graphs. *J. Graph Theory* **2017**, *84*, 158–175. [CrossRef]
10. Máčajová, E.; Mazák, J. On even cycle decompositions of 4-regular line graphs. *Discrete Math.* **2013**, *313*, 1697–1699. [CrossRef]
11. Markström, K. Even cycle decompositions of 4-regular graphs and line graphs. *Discrete Math.* **2012**, *312*, 2676–2681. [CrossRef]
12. Seymour, P.D. Even circuits in planar graphs. *J. Comb. Theory Ser. B* **1981**, *31*, 327–338. [CrossRef]
13. Thomassen, C. Even cycles in directed graphs. *Eur. J. Comb.* **1985**, *6*, 85–89. [CrossRef]
14. Yuster, R.; Zwick, U. Finding even cycles even faster. *SIAM J. Discrete Math.* **1997**, *10*, 209–222. [CrossRef]
15. Zhang, C.-Q. On even circuit decompositions of Eulerian graphs. *J. Graph Theory* **1994**, *18*, 51–57. [CrossRef]
16. Francisco, C.A.; Hà, H.T.; van Tuyl, A. Associated primes of monomial ideals and odd holes in graphs. *J. Algebr. Comb.* **2010**, *32*, 287–301. [CrossRef]
17. Francisco, C.A.; Hà, H.T.; Mermin, J. *Powers of Square-Free Monomial Ideals and Combinatorics*; Commutative Algebra; Springer: New York, NY, USA, 2013; pp. 373–392.
18. Morey, S.; Villarreal, R.H. *Edge Ideals: Algebraic and Combinatorial Properties*; Progress in commutative algebra 1; de Gruyter: Berlin, Germany, 2012; pp. 85–126.
19. Villarreal, R.H. Cohen-Macaulay graphs. *Manuscr. Math.* **1990**, *66*, 277–293. [CrossRef]
20. Faridi, S. The facet ideal of a simplicial complex. *Manuscr. Math.* **2002**, *109*, 159–174. [CrossRef]
21. Hà, H.T.; van Tuyl, A. Monomial ideals, edge ideals of hypergraphs, and their graded Betti numbers. *J. Algebr. Comb.* **2008**, *27*, 215–245. [CrossRef]
22. Dulmage, A.L.; Mendelsohn, N.S. Coverings of bipartite graphs. *Can. J. Math.* **1958**, *10*, 517–534. [CrossRef]
23. Kundu, S.; Lawler, E. A matroid generalization of a theorem of Mendelsohn and Dulmage. *Discrete Math.* **1973**, *4*, 159–163. [CrossRef]
24. Herzog, J.; Hibi, T. *Monomial Ideals*; Graduate Texts in Mathematics 260; Springer: Berlin, Germany, 2011.

25. Villarreal, R.H. *Monomial Algebras*, 2nd ed.; Monographs and Research Notes in Mathematics; CRC Press: Boca Raton, FL, USA, 2015; pp. xviii + 686, ISBN 978-1-4822-3469-5.
26. Diestel, R. *Graph Theory*, 5th ed.; Graduate Texts in Mathematics, 173; Springer: Berlin, Germany, 2018; pp. xviii + 428, ISBN 978-3-662-57560-4, 978-3-662-53621-6.
27. Vasconcelos, W.V. *Arithmetic of Blowup Algebras*; London Math. Soc., Lecture Note Series 195; Cambridge University Press: Cambridge, UK, 1994.
28. Morey, S. Equations of blowups of ideals of codimension two and three. *J. Pure Appl. Algebra* **1996**, *109*, 197–211. [CrossRef]
29. Johnson, M.; Morey, S. Normal blow-ups and their expected defining equations. *J. Pure Appl. Algebra* **2001**, *162*, 303–313. [CrossRef]
30. Kustin, A.; Polini, C.; Ulrich, B. The equations defining blowup algebras of height three Gorenstein ideals. *Algebra Number Theory* **2017**, *11*, 1489–1525. [CrossRef]
31. Ulrich, B.; Vasconcelos, W.V. The equations of Rees algebras of ideals with linear presentation. *Math. Z.* **1993**, *214*, 79–92. [CrossRef]
32. Vasconcelos, W.V. On the equations of Rees algebras. *J. Reine Angew. Math.* **1991**, *418*, 189–218.
33. Bayer, D.; Peeva, I.; Sturmfels, B. Monomial resolutions. *Math. Res. Lett.* **1998**, *5*, 31–46. [CrossRef]
34. Taylor, D. Ideals Generated by Monomials in an *R*-Sequence. Ph.D. Thesis, University of Chicago, Chicago, IL, USA, 1966.
35. Villarreal, R.H. Rees algebras of edge ideals. *Commun Algebra* **1995**, *23*, 3513–3524. [CrossRef]
36. Villarreal, R.H. (Departamento de Matemáticas, Centro de Investigación y de Estudios Avanzados del IPN, Apartado Postal 14–740, 07000 Mexico City, D.F., Mexico). Personal Communication, 2019.
37. Caccetta, L.; Häggkvist, R. On minimal digraphs with given girth. In Proceedings of the Ninth Southeastern Conference on Combinatorics, Graph Theory, and Computing, Boca Raton, FL, USA, 30 January–2 February 1978; pp. 181–187.

© 2019 by the authors. Licensee MDPI, Basel, Switzerland. This article is an open access article distributed under the terms and conditions of the Creative Commons Attribution (CC BY) license (http://creativecommons.org/licenses/by/4.0/).

Article

# The Regularity of Edge Rings and Matching Numbers

Jürgen Herzog [1] and Takayuki Hibi [2],*

[1] Fachbereich Mathematik, Universität Duisburg-Essen, Campus Essen, 45117 Essen, Germany; juergen.herzog@uni-essen.de
[2] Department of Pure and Applied Mathematics, Graduate School of Information Science and Technology, Osaka University, Suita, Osaka 565-0871, Japan
\* Correspondence: hibi@math.sci.osaka-u.ac.jp

Received: 30 November 2019; Accepted: 18 December 2019; Published: 1 January 2020

**Abstract:** Let $K[G]$ denote the edge ring of a finite connected simple graph $G$ on $[d]$ and $\mathrm{mat}(G)$ the matching number of $G$. It is shown that $\mathrm{reg}(K[G]) \leq \mathrm{mat}(G)$ if $G$ is non-bipartite and $K[G]$ is normal, and that $\mathrm{reg}(K[G]) \leq \mathrm{mat}(G) - 1$ if $G$ is bipartite.

**Keywords:** edge ring; edge polytope; regularity; matching number

---

Let $G$ be a finite connected simple graph on the vertex set $[d] = \{1, \ldots, d\}$ and let $E(G)$ be its edge set. Let $S = K[x_1, \ldots, x_d]$ denote the polynomial ring in $d$ variables over a field $K$. The *edge ring* of $G$ is the toric ring $K[G] \subset S$ which is generated by those monomials $x_i x_j$ with $\{i, j\} \in E(G)$. The systematic study of edge rings originated in [1]. It has been shown that $K[G]$ is normal if and only if $G$ satisfies the odd cycle condition ([2], p. 131). Thus, particularly if $G$ is bipartite, $K[G]$ is normal.

Let $\mathbf{e}_1, \ldots, \mathbf{e}_d$ denote the canonical unit coordinate vectors of $\mathbb{R}^d$. The *edge polytope* is the lattice polytope $\mathcal{P}_G \subset \mathbb{R}^d$ which is the convex hull of the finite set $\{ \mathbf{e}_i + \mathbf{e}_j : \{i, j\} \in E(G) \}$. One has $\dim \mathcal{P}_G = d - 1$ if $G$ is non-bipartite and $\dim \mathcal{P}_G = d - 2$ if $G$ is bipartite. We refer the reader to ([2], Chapter 5) for the fundamental materials on edge rings and edge polytopes.

A *matching* of $G$ is a subset $M \subset E(G)$ for which $e \cap e' = \emptyset$ for $e \neq e'$ belonging to $M$. The *matching number* is the maximal cardinality of matchings of $G$. Let $\mathrm{mat}(G)$ denote the matching number of $G$.

When $K[G]$ is normal, the upper bound of regularity of $K[G]$ can be explicitly described in terms of $\mathrm{mat}(G)$. Our main result in the present paper is as follows:

**Theorem 1.** *Let $G$ be a finite connected simple graph. Then*

(a) *If $G$ is non-bipartite and $K[G]$ is normal, then $\mathrm{reg}\, K[G] \leq \mathrm{mat}(G)$;*
(b) *If $G$ is bipartite, then $\mathrm{reg}\, K[G] \leq \mathrm{mat}(G) - 1$.*

Lemma 1 stated below, which provides information on lattice points belonging to the interiors of dilations of edge polytopes, is indispensable for the proof of Theorem 1.

**Lemma 1.** *Suppose that $(a_1, \ldots, a_d) \in \mathbb{Z}^d$ belongs to the interior $q(\mathcal{P}_G \setminus \partial \mathcal{P}_G)$ of the dilation $q\mathcal{P}_G = \{q\alpha : \alpha \in \mathcal{P}_G\}$, where $q \geq 1$, of $\mathcal{P}_G$. Then $a_i \geq 1$ for each $1 \leq i \leq d$.*

**Proof.** The facets of $\mathcal{P}_G$ are described in ([1], Theorem 1.7). When $W \subset [d]$, we write $G_W$ for the induced subgraph of $G$ on $W$. Since $K[G]$ is normal, it follows that $\mathcal{P}_G$ possesses the integer decomposition property ([2], p. 91). In other words, each $\mathbf{a} \in q\mathcal{P}_G \cap \mathbb{Z}^d$ is of the form

$$\mathbf{a} = (\mathbf{e}_{i_1} + \mathbf{e}_{j_1}) + \cdots + (\mathbf{e}_{i_q} + \mathbf{e}_{j_q}),$$

where $\{i_1, j_1\}, \ldots, \{i_q, j_q\}$ are edges of $G$.

**(First Step)** Let $G$ be non-bipartite. Let $i \in [d]$. Let $H_1, \ldots, H_s$ and $H'_1, \ldots, H'_{s'}$ denote the connected components of $G_{[d]\setminus\{i\}}$, where each $H_j$ is bipartite and where each $H'_{j'}$ is non-bipartite. If $s = 0$, then $i \in [d]$ is regular ([1], p. 414) and the hyperplane of $\mathbb{R}^d$ defined by the equation $x_i = 0$ is a facet of $q\mathcal{P}_G$. Hence $a_i > 0$.

Let $s \geq 1$ and $s' \geq 0$. For each $1 \leq j \leq s$, we write $W_j \cup U_j$ for the vertex set of the bipartite graph $H_j$ for which there is $a \in W_j$ with $\{a, i\} \in E(G)$, where $U_j = \emptyset$ if $H_j$ is a graph consisting of a single vertex. Then $T = W_1 \cup \cdots \cup W_s$ is independent ([1], p. 414). In other words, no edge $e \in E(G)$ satisfies $e \subset T$. Let $G'$ denote the bipartite graph induced by $T$. Thus the edges of $G'$ are $\{b, c\} \in E(G)$ with $b \in T$ and $c \in T' = U_1 \cup \cdots \cup U_s \cup \{i\}$. Since each induced subgraph $G_{W_j \cup U_j \cup \{i\}}$ is connected, it follows that $G'$ is connected with $V(G') = T \cup T'$ as its vertex set. Since the connected components of $G_{[d]\setminus V(G')}$ are $H'_1, \ldots, H'_{s'}$, it follows that $T$ is fundamental ([1], p. 415) and the hyperplane of $\mathbb{R}^d$ defined by $\sum_{\xi \in T} x_\xi = \sum_{\xi' \in T'} x_{\xi'}$ is a facet of $q\mathcal{P}_G$. Now, suppose that $a_i = 0$. Since $\mathcal{P}_G$ possesses the integer decomposition property, one has $\sum_{\xi \in T} a_\xi = \sum_{\xi' \in T'} a_{\xi'}$. Hence $(a_1, \ldots, a_d) \in \mathbb{Z}^d$ cannot belong to $q(\mathcal{P}_G \setminus \partial \mathcal{P}_G)$. Thus $a_i > 0$, as desired.

**(Second Step)** Let $G$ be bipartite. If $G$ is a star graph with, say, $E(G) = \{\{1, 2\}, \{1, 3\}, \ldots, \{1, d\}\}$, then $\mathcal{P}_G$ can be regarded to be the $(d - 2)$ simplex of $\mathbb{R}^{d-1}$ with the vertices $(1, 0, \ldots, 0), (0, 1, 0, \ldots, 0), \ldots, (0, \ldots, 0, 1)$. Thus, since each $(a_1, \ldots, a_d) \in q\mathcal{P}_G \cap \mathbb{Z}^d$ satisfies $a_1 = q$, the assertion follows immediately. In the argument below, one will assume that $G$ is not a star graph.

Let $i \in [d]$ and $H_1, \ldots, H_s$ be the connected components of $G_{[d]\setminus\{i\}}$. If $s = 1$, then $i \in [d]$ is ordinary ([1], p. 414) and the hyperplane of $\mathbb{R}^d$ defined by the equation $x_i = 0$ is a facet of $q\mathcal{P}_G$. Hence $a_i > 0$.

Let $s \geq 2$. Let $W_j \cup U_j$ denote the vertex set of $H_j$ for which there is $a \in W_j$ with $\{a, i\} \in E(G)$. Since $G$ is not a star graph, one can assume that $U_1 \neq \emptyset$. Then $T = W_2 \cup \cdots \cup W_s$ is independent and the bipartite graph induced by $T$ is $G_{[d]\setminus(W_1 \cup U_1)}$. Hence $T$ is acceptable ([1], p. 415) and the hyperplane of $\mathbb{R}^d$ defined by $\sum_{\xi \in W_1} x_\xi = \sum_{\xi' \in U_1} x_{\xi'}$ is a facet of $q\mathcal{P}_G$. Now, suppose that $a_i = 0$. Since $\mathcal{P}_G$ possesses the integer decomposition property, one has $\sum_{\xi \in W_1} a_\xi = \sum_{\xi' \in U_1} a_{\xi'}$. Hence $(a_1, \ldots, a_d) \in \mathbb{Z}^d$ cannot belong to $q(\mathcal{P}_G \setminus \partial \mathcal{P}_G)$. Thus $a_i > 0$, as required. □

We say that a finite subset $L \subset E(G)$ is an *edge cover* of $G$ if $\cup_{e \in L} e = [d]$. Let $\mu(G)$ denote the minimal cardinality of edge covers of $G$.

**Corollary 1.** *When $K[G]$ is normal, one has $q \geq \mu(G)$ if $q(\mathcal{P}_G \setminus \partial \mathcal{P}_G) \cap \mathbb{Z}^d \neq \emptyset$.*

**Proof.** Since $\mathcal{P}_G$ possesses the integer decomposition property, Lemma 1 guarantees that, if $\mathbf{a} \in q(\mathcal{P}_G \setminus \partial \mathcal{P}_G) \cap \mathbb{Z}^d$, one has $q \geq \mu(G)$. □

Once Corollary 1 is established, to complete the proof of Theorem 1 is a routine job on computing the regularity of normal toric rings.

**Proof of Theorem 1.** In each of the cases (a) and (b), since the edge ring $K[G]$ is normal, it follows that the Hilbert function of $K[G]$ coincides the Ehrhart function ([2], p. 100) of the edge polytope $\mathcal{P}_G$, which says that the Hilbert series of $K[G]$ is of the form

$$(h_0 + h_1 \lambda + \cdots + h_s \lambda^s)/(1 - \lambda)^{(\dim \mathcal{P}_G) + 1}$$

with each $h_i \in \mathbb{Z}$ and $h_s \neq 0$. One has

$$s = (\dim \mathcal{P}_G + 1) - \min\{q \geq 1 : q(\mathcal{P}_G \setminus \partial \mathcal{P}_G) \cap \mathbb{Z}^d \neq \emptyset\}.$$

Now, Corollary 1 guarantees that
$$s \leq (\dim \mathcal{P}_G + 1) - \mu(G).$$

Finally, since $\mu(G) = d - \mathrm{mat}(G)$ ([3], Lemma 2.1), one has
$$\mathrm{reg}\, K[G] = s \leq \dim \mathcal{P}_G - (d-1) + \mathrm{mat}(G),$$

as required. □

Rafael H. Villarreal informed us that part (b) of Theorem 1 can also be deduced from ([4], Theorem 14.4.19).

When $K[G]$ is non-normal, the behavior of regularity is curious.

**Proposition 1.** *For given integers $0 \leq r \leq m$, there exists a finite connected simple graph $G$ such that $\mathrm{reg}\, K[G] = r$, and*
$$\mathrm{mat}(G) = \begin{cases} m, & \text{if } G \text{ is non-bipartite,} \\ m+1, & \text{if } G \text{ is bipartite.} \end{cases}$$

**Proof.** In the non-bipartite case, let $H$ be the complete graph with $2r$ vertices. Its matching number is $r$. We know from ([5], Corollary 2.12) that $\mathrm{reg}\, K[H] = r$. At one vertex of $H$ we attach a path graph of length $2(m-r)$ and call this new graph $G$. Then $\mathrm{mat}(G) = m$ and $\mathrm{reg}\, K[G] = \mathrm{reg}\, K[H] = r$, as $K[G]$ is just a polynomial extension of $K[H]$.

In the bipartite case, let $H$ be the bipartite graph of type $(r+1, r+1)$. The matching number is $r+1$. Indeed, $K[H]$ may be viewed as a Hibi ring whose regularity is well-known, see for example ([6], Theorem 1.1). At one vertex of $H$ we attach a path graph of length $2(m-r)$ and call this new graph $G$. Then $\mathrm{mat}(G) = m+1$ and $\mathrm{reg}\, K[G] = \mathrm{reg}\, K[H] = r$, for the same reason as before. □

These bounds for the regularity of $K[G]$ are generally only valid if $K[G]$ is normal. Consider, for example, the graph $G$ which consists of two disjoint triangles combined as a path of length $\ell$. Then the defining ideal of $K[G]$ is generated by a binomial of degree $\ell + 3$, and hence $\mathrm{reg}\, K[G] = \ell + 2$, while the matching number of $G$ is $2 + \lceil \ell/2 \rceil$.

**Question 1.** *Let $m$ be a positive integer, and consider the set $\mathcal{S}_m$ of finite connected simple graphs with matching number $m$.*

- *Is there a bound for $\mathrm{reg}\, K[G]$ with $G \in \mathcal{S}_m$?*
- *If such a bound exists, is it a linear function of $m$?*

**Author Contributions:** All authors made equal and significant contributions to writing this article, and approved the final manuscript. All authors have read and agreed to the published version of the manuscript.

**Funding:** Takayuki Hibi was partially supported by JSPS KAKENHI 19H00637.

**Conflicts of Interest:** The authors declare no conflict of interest. The funders had no role in the design of the study; in the collection, analyses, or interpretation of data; in the writing of the manuscript, or in the decision to publish the results.

## References

1. Ohsugi, H.; Hibi, T. Normal polytopes arising from finite graphs. *J. Algebra* **1998**, *207*, 409–426. [CrossRef]
2. Herzog, J.; Hibi, T.; Ohsugi, H. *Binomial Ideals*; GTM 279; Springer: New York, NY, USA, 2018.
3. Herzog, J.; Hibi, T. Matching numbers and the regularity of the Rees algebra of an edge ideal. *arXiv* **2019**, arXiv:1905.02141.

4. Villarreal, R.H. *Monomial Algebras*, 2nd ed.; Monographs and Research Notes in Mathematics, Taylor & Francis Group: Abingdon, UK, 2015.
5. Bruns, W.; Vasoncelos, W.V.; Villarreal, R.H. Degree bounds in monomial subrings. *Ill. J. Math.* **1997**, *41*, 341–353. [CrossRef]
6. Ene, V.; Herzog, J.; Madani, S.S. A note on the regularity of Hibi rings. *Manuscripta Math.* **2015**, *148*, 501–506. [CrossRef]

© 2020 by the authors. Licensee MDPI, Basel, Switzerland. This article is an open access article distributed under the terms and conditions of the Creative Commons Attribution (CC BY) license (http://creativecommons.org/licenses/by/4.0/).

Article
# Odd Cycles and Hilbert Functions of Their Toric Rings

Takayuki Hibi [1] and Akiyoshi Tsuchiya [2,*]

[1] Department of Pure and Applied Mathematics, Graduate School of Information Science and Technology, Osaka University, Suita, Osaka 565-0871, Japan; hibi@math.sci.osaka-u.ac.jp
[2] Graduate School of Mathematical Sciences, University of Tokyo, Komaba, Meguro-ku, Tokyo 153-8914, Japan
* Correspondence: akiyoshi@ms.u-tokyo.ac.jp

Received: 3 December 2019; Accepted: 18 December 2019; Published: 20 December 2019

**Abstract:** Studying Hilbert functions of concrete examples of normal toric rings, it is demonstrated that for each $1 \leq s \leq 5$, an O-sequence $(h_0, h_1, \ldots, h_{2s-1}) \in \mathbb{Z}_{\geq 0}^{2s}$ satisfying the properties that (i) $h_0 \leq h_1 \leq \cdots \leq h_{s-1}$, (ii) $h_{2s-1} = h_0$, $h_{2s-2} = h_1$ and (iii) $h_{2s-1-i} = h_i + (-1)^i$, $2 \leq i \leq s-1$, can be the $h$-vector of a Cohen-Macaulay standard G-domain.

**Keywords:** O-sequence; $h$-vector; flawless; toric ring; stable set polytope

**MSC:** 13A02; 13H10

## 1. Background

In the paper [1] published in 1989, several conjectures on Hilbert functions of Cohen-Macaulay integral domains are studied.

Let $A = \bigoplus_{n=0}^{\infty} A_n$ be a standard G-algebra [2]. Thus $A$ is a Noetherian commutative graded ring for which (i) $A_0 = K$ a field, (ii) $A = K[A_1]$ and (iii) $\dim_K A_1 < \infty$. The Hilbert function of $A$ is defined by

$$H(A, n) = \dim_K A_n, \quad n = 0, 1, 2, \ldots$$

Let $\dim A = d$ and $v = H(A, 1) = \dim_K A_1$. A classical result ([3], Chapter 5, Section 13) says that $H(A, n)$ is a polynomial for $n$ sufficiently large and its degree is $d - 1$. It follows that the sequence $h(A) = (h_0, h_1, h_2, \ldots)$, called the $h$-vector of $A$, defined by the formula

$$(1 - \lambda)^d \sum_{n=0}^{\infty} H(A, n) \lambda^n = \sum_{i=0}^{\infty} h_i \lambda^i$$

has finitely many non-zero terms with $h_0 = 1$ and $h_1 = v - d$. If $h_i = 0$ for $i > s$ and $h_s \neq 0$, then we write $h(A) = (h_0, h_1, \ldots, h_s)$.

Let $Y_1, \ldots, Y_r$ be indeterminates. A non-empty set $M$ of monomials $Y_1^{a_1} \cdots Y_r^{a_r}$ in the variables $Y_1, \ldots, Y_r$ is said to be an order ideal of monomials if, whenever $m \in M$ and $m'$ divides $m$, then $m' \in M$. Equivalently, if $Y_1^{a_1} \cdots Y_r^{a_r} \in M$ and $0 \leq b_i \leq a_i$, then $Y_1^{b_1} \cdots Y_r^{b_r} \in M$. In particular, since $M$ is non-empty, $1 \in M$. A finite sequence $(h_0, h_1, \ldots, h_s)$ of non-negative integers is said to be an O-sequence if there exists an order ideal $M$ of monomials in $Y_1, \ldots, Y_r$ with each $\deg Y_i = 1$ such that $h_j = |\{m \in M | \deg m = j\}|$ for any $0 \leq j \leq s$. In particular, $h_0 = 1$. If $A$ is Cohen-Macaulay, then $h(A) = (h_0, h_1, \ldots, h_s)$ is an O-sequence ([2], p. 60). Furthermore, a finite sequence $(h_0, h_1, \ldots, h_s)$ of integers with $h_0 = 1$ and $h_s \neq 0$ is the $h$-vector of a Cohen-Macaulay standard $G$-algebra if and only if $(h_0, h_1, \ldots, h_s)$ is an $O$-sequence ([2], Corollary 3.11).

An $O$-sequence $(h_0, h_1, \ldots, h_s)$ with $h_s \neq 0$ is called *flawless* ([1], p. 245) if (i) $h_i \leq h_{s-i}$ for $0 \leq i \leq [s/2]$ and (ii) $h_0 \leq h_1 \leq \cdots \leq h_{[s/2]}$. A standard $G$-domain is a standard $G$-algebra which is an integral domain. It was conjectured ([1], Conjecture 1.4) that the $h$-vector of a Cohen-Macaulay standard $G$-domain is flawless. Niesi and Robbiano ([4], Example 2.4) succeeded in constructing a Cohen-Macaulay standard $G$-domain with $(1, 3, 5, 4, 4, 1)$ its $h$-vector. Thus, in general, the $h$-vector of a Cohen-Macaulay standard $G$-domain is not flawless.

In the present paper, it is shown that, for each $1 \leq s \leq 5$, an $O$-sequence

$$(h_0, h_1, \ldots, h_{s-1}, h_s, \ldots, h_{2s-2}, h_{2s-1}) \in \mathbb{Z}_{\geq 0}^{2s}$$

satisfying the properties that

(i) $h_0 \leq h_1 \leq \cdots \leq h_{s-1}$,
(ii) $h_{2s-1} = h_0$, $h_{2s-2} = h_1$,
(iii) $h_{2s-1-i} = h_i + (-1)^i$, $2 \leq i \leq s-1$

can be the $h$-vector of a normal toric ring arising from a cycle of odd length. In particular, the above $O$-sequence, which is non-flawless for each of $s = 4$ and $s = 5$, can be the $h$-vector of a Cohen-Macaulay standard $G$-domain.

## 2. Toric Rings Arising from Odd Cycles

Let $C_{2s+1}$ denote a cycle of length $2s+1$, where $s \geq 1$, on $[2s+1] = \{1, 2, \ldots, 2s+1\}$ with the edges

$$\{1, 2\}, \{2, 3\}, \ldots, \{2s-1, 2s\}, \{2s, 2s+1\}, \{2s+1, 1\}. \tag{1}$$

A finite set $W \subset [2s+1]$ is called *stable* in $C_{2s+1}$ if none of the sets of (1) is a subset of $W$. In particular, the empty set $\emptyset$ and $\{1\}, \{2\}, \ldots, \{2s+1\}$ are stable. Let $S = K[x_1, \ldots, x_{2s+1}, y]$ denote the polynomial ring in $2s+2$ variables over $K$. The *toric ring* of $C_{2s+1}$ is the subring $K[C_{2s+1}]$ of $S$ which is generated by those squarefree monomials $(\prod_{i \in W} x_i) y$ for which $W \subset [2s+1]$ is stable in $C_{2s+1}$. It follows that $K[C_{2s+1}]$ can be a standard $G$-algebra with each $\deg(\prod_{i \in W} x_i) y = 1$. It is shown ([5], Theorem 8.1) that $K[C_{2s+1}]$ is normal. In particular, $K[C_{2s+1}]$ is a Cohen-Macaulay standard $G$-domain. Now, we discuss when $K[C_{2s+1}]$ is Gorenstein. Here a Cohen-Macaulay ring is called Gorenstein if it has finite injective dimension.

**Theorem 1.** *The toric ring $K[C_{2s+1}]$ is Gorenstein if and only if either $s = 1$ or $s = 2$.*

**Proof.** Since the $h$-vector of $K[C_3]$ is $(1, 1)$ and since the $h$-vector of $K[C_5]$ is $(1, 6, 6, 1)$, it follows from ([2], Theorem 4.4) that each of $K[C_3]$ and $K[C_5]$ is Gorenstein.

Now, we show that $K[C_{2s+1}]$ is not Gorenstein if $s \geq 3$. Let $s \geq 3$. Write $\mathcal{Q}_{C_{2s+1}} \subset \mathbb{R}^{2s+1}$ for the stable set polytope of $C_{2s+1}$. Thus $\mathcal{Q}_{C_{2s+1}}$ is the convex hull of the finite set

$$\left\{ \sum_{i \in W} \mathbf{e}_i : W \text{ is a stable set of } G \right\} \subset \mathbb{R}^{2s+1},$$

where $\mathbf{e}_1, \ldots, \mathbf{e}_{2s+1} \in \mathbb{R}^{2s+1}$ are the canonical unit coordinate vectors of $\mathbb{R}^{2s+1}$ and where $\sum_{i \in \emptyset} \mathbf{e}_i = (0, \ldots, 0) \in \mathbb{R}^{2s+1}$. One has $\dim \mathcal{Q}_{2s+1} = 2s+1$. Then ([6], Theorem 4) says that $\mathcal{Q}_{C_{2s+1}}$ is defined by the following inequalities:

- $0 \leq x_i \leq 1$ for all $1 \leq i \leq 2s+1$;
- $x_i + x_{i+1} \leq 1$ for all $1 \leq i \leq 2s$;
- $x_1 + x_{2s+1} \leq 1$;
- $x_1 + \cdots + x_{2s+1} \leq s$.

It then follows that each of $\mathcal{Q}_{C_{2s+1}}$ and $2\mathcal{Q}_{C_{2s+1}}$ has no interior lattice points and that $(1,\ldots,1)$ is an interior lattice point of $3\mathcal{Q}_{C_{2s+1}}$. Furthermore, (Ref. [7], Theorem 4.2) guarantees that the inequality

$$x_1 + \cdots + x_{2s+1} \leq s$$

defines a facet of $\mathcal{Q}_{C_{2s+1}}$. Let $\mathcal{P}_s = 3\mathcal{Q}_{C_{2s+1}} - (1,\ldots,1)$. Thus the origin of $\mathbb{R}^{2s+1}$ is an interior lattice point of $\mathcal{P}_s$ and the inequality

$$x_1 + \cdots + x_{2s+1} \leq s - 1$$

defines a facet of $\mathcal{P}_s$. This fact together with [8] implies that $\mathcal{P}_s$ is not reflexive. In other words, the dual polytope $\mathcal{P}_s^\vee$ of $\mathcal{P}_s$ defined by

$$\mathcal{P}_s^\vee = \{ \mathbf{y} \in \mathbb{R}^{2s+1} : \langle \mathbf{x}, \mathbf{y} \rangle \leq 1 \text{ for all } \mathbf{x} \in \mathcal{P}_s \}$$

is not a lattice polytope, where $\langle \mathbf{x}, \mathbf{y} \rangle$ is the usual inner product of $\mathbb{R}^{2s+1}$. It then follows from ([9], Theorem (1.1)) (and also from ([5], Theorem 8.1)) that $K[C_{2s+1}]$ is not Gorenstein, as desired. □

It is known ([2], Theorem 4.4) that a Cohen-Macaulay standard $G$-domain $A$ is Gorenstein if and only if the $h$-vector $h(A) = (h_0, \ldots, h_s)$ is symmetric, i.e., $h_i = h_{s-i}$ for $0 \leq i \leq [s/2]$. Hence the $h$-vector of the toric ring $K[C_{2s+1}]$ is not symmetric when $s \geq 3$.

**Example 1.** *By using Normaliz [10], the h-vector of the toric ring $K[C_7]$ is $(1, 21, 84, 85, 21, 1)$.*

## 3. Non-Flawless O-Sequences of Normal Toric Rings

We now come to concrete examples of non-flawless O-sequences which can be the $h$-vectors of normal toric rings.

**Example 2.** *The h-vector of the toric ring $K[C_9]$ is*

$$(1, 66, 744, 2305, 2304, 745, 66, 1).$$

*Furthermore,*

$$(1, 187, 5049, 37247, 96448, 96449, 37246, 5050, 187, 1)$$

*is the h-vector of the toric ring $K[C_{11}]$.*

We conclude the present paper with the following

**Conjecture 1.** *The h-vector of the toric ring $K[C_{2s+1}]$ of $C_{2s+1}$ is of the form*

$$(1, h_1, h_2, h_3, \ldots, h_i, \ldots, h_{s-1}, h_{s-1} + (-1)^{s-1}, \ldots, h_i + (-1)^i, \ldots, h_3 - 1, h_2 + 1, h_1, 1).$$

**Author Contributions:** All authors made equal and significant contributions to writing this article, and approved the final manuscript. All authors have read and agreed to the published version of the manuscript.

**Funding:** Takayuki Hibi was partially supported by JSPS KAKENHI 19H00637. Akiyoshi Tsuchiya was partially supported by JSPS KAKENHI 19K14505 and 19J00312.

**Conflicts of Interest:** The authors declare no conflict of interest. The funders had no role in the design of the study; in the collection, analyses, or interpretation of data; in the writing of the manuscript, or in the decision to publish the results.

## References

1. Hibi, T. Flawless *O*-sequences and Hilbert functions of Cohen-Macaulay integral domains. *J. Pure Appl. Algebra* **1989**, *60*, 245–251. [CrossRef]
2. Stanley, R. Hilbert functions of graded algebras. *Adv. Math.* **1978**, *28*, 57–83. [CrossRef]
3. Matsumura, H. *Commutative Ring Theory*; Cambridge University Press: Cambridge, UK, 1989.
4. Niesi, G.; Robbiano, L. Disproving Hibi's Conjecture with CoCoA or Projective Curves with bad Hilbert Functions. In *Computational Algebraic Geometry*; Eyssette, F., Galligo, A., Eds.; Birkhäuser: Boston, MA, USA, 1993; pp. 195–201.
5. Engström, A.; Norén, P. Ideals of Graphs Homomorphisms. *Ann. Comb.* **2013**, *17*, 71–103. [CrossRef]
6. Mahjoub, A.R. On the stable set polytope of a series-parallel graph. *Math. Programm.* **1988**, *40*, 53–57. [CrossRef]
7. Chvátal, V. On certain polytopes associated with graphs. *J. Comb. Theory Ser. B* **1975**, *18*, 138–154. [CrossRef]
8. Hibi, T. Dual polytopes of rational convex polytopes. *Combinatorica* **1992**, *12*, 237–240. [CrossRef]
9. De Negri, E.; Hibi, T. Gorenstein algebras of Veronese type. *J. Algebra* **1997**, *193*, 629–639. [CrossRef]
10. Bruns, W.; Ichim, B.; Römer, T.; Sieg, R.; Söger, C. Normaliz, Algorithms for Rational Cones and Affine Monoids. Available online: https://www.normaliz.uni-osnabrueck.de (accessed on 1 December 2019).

© 2019 by the authors. Licensee MDPI, Basel, Switzerland. This article is an open access article distributed under the terms and conditions of the Creative Commons Attribution (CC BY) license (http://creativecommons.org/licenses/by/4.0/).

*Article*

# Cohen-Macaulay and ($S_2$) Properties of the Second Power of Squarefree Monomial Ideals

**Do Trong Hoang [1,†], Giancarlo Rinaldo [2,†] and Naoki Terai [3,\*,†]**

[1] Institute of Mathematics, Vietnam Academy of Science and Technology, 18 Hoang Quoc Viet, Hanoi 10307, Vietnam
[2] Department of Mathematics, University of Trento, via Sommarive, 14, 38123 Povo (Trento), Italy
[3] Faculty of Education, Saga University, Saga 840-8502, Japan
\* Correspondence: terai@cc.saga-u.ac.jp
† These authors contributed equally to this work.

Received: 16 June 2019; Accepted: 30 July 2019; Published: 31 July 2019

**Abstract:** We show that Cohen-Macaulay and ($S_2$) properties are equivalent for the second power of an edge ideal. We give an example of a Gorenstein squarefree monomial ideal $I$ such that $S/I^2$ satisfies the Serre condition ($S_2$), but is not Cohen-Macaulay.

**Keywords:** Stanley-Reisner ideal; edge ideal; Cohen-Macaulay; ($S_2$) condition

## 1. Introduction

Let $K$ be a fixed field. Let $S = K[x_1, \ldots, x_n]$ be a polynomial ring with $\deg x_i = 1$ for all $i \in [n] = \{1, 2, \ldots, n\}$. Let $I$ be a squarefree monomial ideal.

For a Stanley-Reisner ring $S/I$, the Cohen-Macaulay and ($S_2$) properties are different in general. For instance, consider the Stanley-Reisner ring of a non-Cohen-Macaulay manifold, e.g., a torus, which satisfies the ($S_2$) condition. However, for some special classes of such rings, they are known to be equivalent. The quotient ring of the edge ideal of a very well-covered graph (see [1]) and a Stanley-Reisner ring with "large" multiplicity (see [2] for the precise statement) are such examples. What about the powers of squarefree monomial ideals?

As for the third and larger powers, the following is proven in [3]:

**Theorem 1.** *Let $I$ be a squarefree monomial ideal. Then, the following conditions are equivalent for a fixed integer $m \geq 3$:*

1. *$S/I$ is a complete intersection.*
2. *$S/I^m$ is Cohen-Macaulay.*
3. *$S/I^m$ satisfies the Serre condition ($S_2$).*

Then, what about the second power of a squarefree monomial ideal? This is the theme of this article. If the second power $I^2$ is Cohen-Macaulay, $I$ is not necessarily a complete intersection. Gorenstein ideals with height three give such examples.

In Section 3, we prove that the Cohen-Macaulay and ($S_2$) properties are equivalent for the second power of a squarefree monomial ideal generated in degree two.

**Theorem 2.** *Let $I$ be a squarefree monomial ideal generated in degree two. Then, the following conditions are equivalent:*

1. *$S/I^2$ is Cohen-Macaulay.*

2. $S/I^2$ satisfies the Serre condition $(S_2)$.

In Section 4, we first give an upper bound of the number of variables in terms of the dimension of $S/I$ when $I$ is a squarefree monomial ideal generated in degree two and $S/I^2$ has the Cohen-Macaulay (equivalently $(S_2)$) property. Using a computer, we classify squarefree monomial ideals $I$ generated in degree two with $\dim S/I \leq 4$ such that $S/I^2$ have the Cohen-Macaulay (equivalently $(S_2)$) property. Since not many examples of squarefree monomial ideals $I$ generated in degree two such that $S/I^2$ are Cohen-Macaulay are known, new examples might be useful. See [4,5] for the two- and three-dimensional cases, respectively, and [6,7] for the higher dimensional case. See also [6,8] for the fact that for a very well-covered graph $G$, the second power $I(G)^2$ is not Cohen-Macaulay if the edge ideal $I(G)$ of $G$ is not a complete intersection.

In Section 5, we give an example of a Gorenstein squarefree monomial ideal $I$ such that $S/I^2$ satisfies the Serre condition $(S_2)$, but is not Cohen-Macaulay. Hence, the Cohen-Macaulay and $(S_2)$ properties are different for the second power in general.

## 2. Preliminaries

### 2.1. Stanley-Reisner Ideals

We recall some notation on simplicial complexes and their Stanley-Reisner ideals. We refer the reader to [9–11] for the detailed information.

Set $V = [n] = \{1, 2, \ldots, n\}$. A nonempty subset $\Delta$ of the power set $2^V$ of $V$ is called a *simplicial complex* on $V$ if the following two conditions are satisfied: (i) $\{v\} \in \Delta$ for all $v \in V$, and (ii) $F \in \Delta$, $H \subseteq F$ imply $H \in \Delta$. An element $F \in \Delta$ is called a *face* of $\Delta$. The dimension of $F$, denoted by $\dim F$, is defined by $\dim F = |F| - 1$. The dimension of $\Delta$ is defined by $\dim \Delta = \max\{\dim F : F \in \Delta\}$. We call a maximal face of $\Delta$ a *facet* of $\Delta$. Let $\mathcal{F}(\Delta)$ denote the set of all facets of $\Delta$. We call $\Delta$ *pure* if all its facets have the same dimension. We call $\Delta$ *connected* if for any pair $(p,q)$, $p \neq q$, of vertices of $\Delta$, there is a chain $p = p_0, p_1, p_2, \ldots, p_k = q$ of vertices of $\Delta$ such that $\{p_{i-1}, p_i\} \in \Delta$ for $i = 1, 2, \ldots, k$.

The *Stanley-Reisner ideal* $I_\Delta$ of $\Delta$ is defined by:

$$I_\Delta = (x_{i_1} x_{i_2} \cdots x_{i_p} : 1 \leq i_1 < \cdots < i_p \leq n, \{x_{i_1}, \ldots, x_{i_p}\} \notin \Delta).$$

The quotient ring $K[\Delta] = K[x_1, \ldots, x_n]/I_\Delta$ is called the *Stanley-Reisner ring* of $\Delta$.

We say that $\Delta$ is a Cohen-Macaulay (resp. Gorenstein) complex if $K[\Delta]$ is a Cohen-Macaulay (resp. Gorenstein) ring. A Gorenstein complex $\Delta$ is called *Gorenstein** if $x_i$ divides some minimal monomial generator of $I_\Delta$ for each $i$.

For a face $F \in \Delta$, the *link* and *star* of $F$ are defined by:

$$\begin{aligned} \operatorname{link}_\Delta F &= \{H \in \Delta : H \cup F \in \Delta, H \cap F = \emptyset\}, \\ \operatorname{star}_\Delta F &= \{H \in \Delta : H \cup F \in \Delta\}. \end{aligned}$$

The Stanley-Reisner ideal $I_\Delta$ of $\Delta$ has the minimal prime decomposition:

$$I_\Delta = \bigcap_{F \in \mathcal{F}(\Delta)} P_F,$$

where $P_F = (x \in [n] \setminus F)$ for each $F \in \mathcal{F}(\Delta)$. We call $I_\Delta$ *unmixed* if all $P_F$ have the same height for $F \in \mathcal{F}(\Delta)$. Note that $\Delta$ is pure if and only if $I_\Delta$ is unmixed. We define the $\ell^{\text{th}}$ symbolic power of $I_\Delta$ by:

$$I_\Delta^{(\ell)} = \bigcap_{F \in \mathcal{F}(\Delta)} P_F^\ell.$$

For a Noetherian ring $A$, the following condition ($S_i$) for $i = 1, 2, \ldots$ is called *Serre's condition*:

$$(S_i) \text{ depth } A_P \geq \min\{\text{height } P, i\} \text{ for all } P \in \text{Spec}(A).$$

See [12] for more information for Stanley-Reisner rings satisfying Serre's condition ($S_i$).

To introduce a characterization of the ($S_2$) property for the second symbolic power of a Stanley-Reisner ideal, we first define the diameter of a simplicial complex. Let $\Delta$ be a connected simplicial complex. For $p, q$ being two vertices of $\Delta$, the *distance* between $p$ and $q$ is the minimal length $k$ of chains $p = p_0, p_1, p_2, \ldots, p_k = q$ of vertices of $\Delta$ such that $\{p_{i-1}, p_i\} \in \Delta$ for $i = 1, 2, \ldots, k$. The *diameter*, denoted by $\text{diam } \Delta$, is the maximal distance between two vertices in $\Delta$. We set $\text{diam } \Delta = \infty$ if $\Delta$ is disconnected. The ($S_2$) property of the second symbolic power of a Stanley-Reisner ideal is characterized as follows:

**Theorem 3.** ([7], Corollary 3.3) *Let $\Delta$ be a pure simplicial complex. Then, the following conditions are equivalent*:

1. $S/I_\Delta^{(2)}$ *satisfies* ($S_2$).
2. $\text{diam}(\text{link}_\Delta F) \leq 2$ *for any face* $F \in \Delta$ *with* $\dim \text{link}_\Delta F \geq 1$.

### 2.2. Edge Ideals

Let $G$ be a graph, which means a finite simple graph, which has no loops and multiple edges. We denote by $V(G)$ (resp. $E(G)$) the set of vertices (resp. edges) of $G$. We call $F \subseteq V(G)$ an *independent set* of $G$ if any $e \in E(G)$ is not contained in $F$. The independence complex $\Delta(G)$ of $G$ is defined by:

$$\Delta(G) = \{F \subset V(G) : e \not\subseteq F \text{ for any } e \in E(G)\},$$

which is a simplicial complex on the vertex set $V(G)$. We define $\alpha(G)$ by:

$$\alpha(G) = \dim \Delta(G) + 1.$$

We define the *neighbor set* $N_G(a)$ of a vertex $a$ of $G$ by:

$$N_G(a) = \{b \in V : ab \in E(G)\}.$$

Set $N_G[a] := \{a\} \cup N_G(a)$, which is called the *closed neighbor set* of a vertex $a$ of $G$. For $S \subseteq V(G)$, we denote by $G \backslash S$ the induced subgraph on the vertex set $V(G) \backslash S$. Set $G_S := G \backslash N_G[S]$, where $N_G[S] := \cup_{x \in S} N_G[x]$. If $S \in \Delta(G)$, then:

$$\text{link}_{\Delta(G)}(S) = \Delta(G_S).$$

See ([11], Lemma 7.4.3). For $ab \in E(G)$, set $G_{ab} := G \backslash (N_G(a) \cup N_G(b))$.

Set $V(G) = \{1, \ldots, n\}$. Then, the *edge ideal* of $G$, denoted by $I(G)$, is a squarefree monomial ideal of $S = K[x_1, \ldots, x_n]$ defined by:

$$I(G) = (x_i x_j : \{x_i, x_j\} \in E(G)).$$

Note that $I(G) = I_{\Delta(G)}$. We call $G$ *well-covered* (or *unmixed*) if $I(G)$ is unmixed.

**Theorem 4** ([13,14]). *Let $G$ be a graph. Then, the following conditions are equivalent*:

1. $G$ *is triangle-free*.
2. $I(G)^{(2)} = I(G)^2$.

**Theorem 5** ([15]). *Let G be a graph. Then, the following conditions are equivalent:*

1. *G is triangle-free, and $I(G)$ is Gorenstein.*
2. *$S/I(G)^2$ is Cohen-Macaulay.*

### 3. The Second Power of Edge Ideals

In this section, we show that the Cohen-Macaulay and $(S_2)$ properties are equivalent for the second power of an edge ideal.

**Lemma 1.** *Let G be a graph with $\alpha(G) \geq 2$. The following conditions are equivalent:*

1. *$S/I(G)^{(2)}$ satisfies the $(S_2)$ property,*
2. *G is a well-covered graph and satisfies $\operatorname{diam}\Delta(G_F) \leq 2$ for all the independent sets F of G such that $|F| \leq \alpha(G) - 2$,*
3. *$G_{ab}$ is well-covered and satisfies $\alpha(G_{ab}) = \alpha(G) - 1$ for all $ab \in E(G)$.*

**Proof.** (1)$\Leftrightarrow$ (2): By [12], Theorem 8.3, $I(G)$ satisfies the $(S_2)$ property if so does $S/I(G)^{(2)}$. Using [12], Corollary 5.4, we obtain that $\Delta(G)$ is pure. This means that $G$ is well-covered, and thus:

$$\operatorname{dim}\operatorname{link}_{\Delta(G)}(F) = \operatorname{dim}\Delta(G) - |F|$$

and $\operatorname{link}_{\Delta(G)}(F) = \Delta(G_F)$. The result is implied by Theorem 3.

(2) $\Rightarrow$ (3): For all $ab \in E(G)$, we have:

$$\alpha(G_{ab}) \leq \alpha(G) - 1.$$

Let $F$ be an independent set of $G_{ab}$. If $|F| < \alpha(G) - 1$, then $|F| \leq \alpha(G) - 2$. Recall that $G_{ab} = G \setminus (N_G(a) \cup N_G(b))$ and $F \subseteq V(G_{ab})$. This implies that $a,b \notin N_G[F]$. Hence, we obtain that $\{a,b\}$ is an edge of $G_F$. In other words, $\{a,b\}$ is not an independent set of $G_F$. By the assumption, $\operatorname{diam}\Delta(G_F) \leq 2$, there is a vertex $c \in V(G_F)$ such that $\{a,c\}, \{c,b\}$ are independent sets of $G_F$. Thus, $ac, bc \notin E(G_F)$. Hence, $c \in V(G_{ab})$. Therefore, $F \cup \{c\}$ is an independent of $G_{ab}$. Then, $G_{ab}$ is well-covered, and moreover, $\alpha(G_{ab}) = \alpha(G) - 1$.

(3) $\Rightarrow$ (2): By [15], Lemma 4.1 (2), $G$ is a well-covered graph. We will prove that $\operatorname{diam}\Delta(G_F) \leq 2$ for all independent set $F$ with $|F| \leq \alpha(G) - 2$ by induction on $\alpha(G)$.

If $\alpha(G) = 2$, then we must prove $\operatorname{diam}\Delta(G) \leq 2$. For all $a,b \in V(G)$, we assume $\{a,b\} \notin \Delta(G)$. Then, $ab \in E(G)$. By the assumption, $\alpha(G_{ab}) = \alpha(G) - 1 = 1 > 0$. Therefore, we can take a vertex $c$ in $G_{ab}$, and thus, $ac, bc \notin E(G)$. Hence, $\{a,c\}, \{b,c\} \in \Delta(G)$. Therefore, we conclude that $\operatorname{diam}\Delta(G) \leq 2$.

Let $\alpha(G) > 2$, and suppose that the assertion is true for all graphs $G'$ with the same structure as $G$ satisfying the condition "$G_{ab}$ is well-covered and satisfies $\alpha(G_{ab}) = \alpha(G) - 1$ for all $ab \in E(G)$" with $\alpha(G') < \alpha(G)$. For all independent set $F$ of $G$ such that $|F| \leq \alpha(G) - 2$, we divide the proof into the following two cases:

**Case 1:** $F = \emptyset$. In this case, we need to prove that $\operatorname{diam}\Delta(G) \leq 2$. In fact, using the same argument as above, we obtain $\operatorname{diam}\Delta(G) \leq 2$.

**Case 2:** $F \neq \emptyset$. Let $x \in F$. Recall that $G$ is a well-covered graph, and thus, we have $\alpha(G_x) = \alpha(G) - 1$. Hence, $|F\setminus\{x\}| = |F| - 1 \leq \alpha(G) - 3 = \alpha(G_x) - 2$. Note that for all $ab \in E(G_x)$, we have that $(G_x)_{ab}$ and $(G_{ab})_x$ are two induced subgraphs of $G$ on vertex set $V(G) \setminus (N_G[x] \cup N_G(a) \cup N_G(b))$. Thus, $(G_x)_{ab} = (G_{ab})_x$. By the assumption and [15], Lemma 4.1 (1), $(G_{ab})_x$ is a well-covered graph with $\alpha((G_{ab})_x) = \alpha(G_{ab}) - 1$. Therefore, $(G_x)_{ab}$ is also a well-covered graph. Moreover,

$$\alpha((G_x)_{ab}) = \alpha((G_{ab})_x) = \alpha(G_{ab}) - 1 = \alpha(G) - 2 = \alpha(G_x) - 1.$$

Thus, $G_x$ has the same structure as $G$ satisfying the condition "$G_{ab}$ is well-covered and satisfies $\alpha(G_{ab}) = \alpha(G) - 1$ for all $ab \in E(G)$" with $\alpha(G_x) < \alpha(G)$. By the induction hypothesis, we obtain $\operatorname{diam} \Delta((G_x)_{F \setminus \{x\}}) \leq 2$. Note that:

$$(G_x)_{F \setminus \{x\}} = G_x \setminus N_G[F \setminus \{x\}] = G \setminus (N_G[x] \cup N_G[F \setminus \{x\}]) = G \setminus (N_G[F]) = G_F.$$

Therefore, $\Delta(G_F) = \Delta((G_x)_{F \setminus \{x\}})$. Therefore, we conclude that $\operatorname{diam} \Delta(G_F) \leq 2$. □

Then, we get the following theorem.

**Theorem 6.** *Let $G$ be a graph. The following conditions are equivalent:*

1. $S/I(G)^2$ satisfies the $(S_2)$ property,
2. $S/I(G)^2$ is Cohen-Macaulay,
3. $G$ is triangle-free, and $G_{ab}$ is a well-covered graph with $\alpha(G_{ab}) = \alpha(G) - 1$ for all $ab \in E(G)$.

**Proof.** By the statements of Conditions (1), (2) and (3), without loss of generality, we can assume that $G$ contains no isolated vertices.

(2) ⇔ (3): By [15], Theorem 4.4, $S/I(G)^2$ is Cohen-Macaulay if and only if $G$ is triangle-free and in $W_2$, which is a well-covered graph such that the removal of any vertex of $G$ leaves a well-covered graph with the same independence number as $G$. By [15], Lemma 4.2, this is equivalent to the condition that $G$ is triangle-free and $G_{ab}$ is a well-covered graph with $\alpha(G_{ab}) = \alpha(G) - 1$ for all $ab \in E(G)$.

(2) ⇒ (1): It is obvious.

(1) ⇒ (3): If $\alpha(G) = 1$, then $G$ is a complete graph. By the assumption, $G$ is one edge. Therefore, the statement holds true. Now, we assume $\alpha(G) \geq 2$. We know that $S/I(G)^2$ satisfies that $(S_2)$ property if and only if $S/I(G)^{(2)}$ satisfies the $(S_2)$ property and $I(G)^2$ has no embedded associated prime, which means $I(G)^2 = I(G)^{(2)}$. By Theorem 4 and Lemma 1, $G$ is triangle-free, and $G_{ab}$ is well-covered with $\alpha(G_{ab}) = \alpha(G) - 1$ for all $ab \in E(G)$. □

**Question.** *If $S/I(G)^{(2)}$ satisfies the $(S_2)$ property, then is it Cohen-Macaulay?*

The question is affirmative if $G$ is a triangle-free graph by Theorems 4 and 6.

## 4. Classification

The purpose of the section is to classify all graphs $G$ such that $S/I(G)^2$ is Cohen-Macaulay with dimension less than five. First, we give an upper bound of the number of vertices of a graph $G$ such that $S/I(G)^2$ is Cohen-Macaulay.

### 4.1. Upper Bound of the Number of Vertices

**Theorem 7** (**Upper bound**)**.** *Let $G$ be a graph with the vertex set $[n]$. Suppose $G$ has no isolate vertex. If $S/I(G)^2$ is $d$-dimensional Cohen-Macaulay, where $d \geq 3$, then we have $n \leq \frac{d^2+3d-2}{2}$.*

**Proof.** We prove this by induction on $d$. For $d = 3$, we have $n \leq 8$ by [5] (see Proposition 3). Set $N(d) = \frac{d^2+3d-2}{2}$. Let $n$ be the number of vertices of $G$ such that $S/I(G)^2$ is $d$-dimensional and Cohen-Macaulay. Let $i \in [n]$. Then, we have $n = |V(\operatorname{star}_{\Delta(G)}\{i\})| + |([n] \setminus V(\operatorname{star}_{\Delta(G)}\{i\})|$. Since $G$ is triangle-free by Theorem 5, an edge among $\{i, p\}$, $\{i, q\}$ and $\{p, q\}$ belongs to $\Delta(G)$ for any $p, q \in ([n] \setminus V(\operatorname{star}_{\Delta(G)}\{i\})$, where $p \neq q$. By the definition of $\operatorname{star}_{\Delta(G)}\{i\}$, we have $\{i, p\}, \{i, q\} \notin \Delta(G)$. Then, we have $\{p, q\} \in \Delta(G)$. By the fact that $I(G)$ is generated in degree two, all minimal non-faces of $\Delta(G)$ have cardinality two. Now, we know that $\{p, q\} \in \Delta(G)$ for any $p, q \in ([n] \setminus V(\operatorname{star}_{\Delta(G)}\{i\})$; hence, we have $[n] \setminus V(\operatorname{star}_{\Delta(G)}\{i\}) \in \Delta(G)$. By the assumption that $S/I(G)^2$ is $d$-dimensional, we have $|[n] \setminus V(\operatorname{star}_{\Delta(G)}\{i\})| \leq d$. Since $\Delta(G)$ is Gorenstein*, so is $\operatorname{link}_{\Delta(G)}\{i\}$ by [10], Theorem

5.1. By Theorem 5, $I^2_{\text{link}_{\Delta(G)}\{i\}}$ is Cohen-Macaulay. Hence, $|V(\text{star}_{\Delta(G)}\{i\})| = |V(\text{link}_{\Delta(G)}\{i\})| + 1 \leq N(d-1) + 1$ by the induction hypothesis. Therefore, $n \leq N(d-1) + d + 1 = \frac{(d-1)^2 + 3(d-1) - 2}{2} + d + 1 = \frac{d^2 + 3d - 2}{2} = N(d)$. □

*4.2. Classification*

In this subsection, we classify all graphs $G$ such that $S/I(G)^2$ is Cohen-Macaulay with dimension less than five.

**Proposition 1.** (One-dimensional case) *Let $G$ be a graph with the vertex set $[n]$. Suppose $G$ has no isolate vertex. Then, $S/I(G)^2$ is one-dimensional Cohen-Macaulay if and only if $n = 2$ and $I(G) = (x_1 x_2)$.*

**Proposition 2** ([4])**.** (Two-dimensional case) *Let $G$ be a graph with the vertex set $[n]$. Suppose $G$ has no isolate vertex. Then, $S/I(G)^2$ is two-dimensional Cohen-Macaulay if and only if $I(G)$ is one of the following up to the permutation of variables:*

1. If $n = 4$, then $(x_1 x_3, x_2 x_4)$.
2. If $n = 5$, then $(x_1 x_3, x_1 x_4, x_2 x_3, x_2 x_5, x_4 x_5)$.

**Proposition 3** ([5])**.** (Three-dimensional case) *Let $G$ be a graph with the vertex set $[n]$. Suppose $G$ has no isolate vertex. Then, $S/I(G)^2$ is three-dimensional Cohen-Macaulay if and only if $I(G)$ is one of the following up to the permutation of variables:*

1. If $n = 6$, then $(x_1 x_4, x_2 x_5, x_3 x_6)$.
2. If $n = 7$, then $(x_1 x_5, x_1 x_6, x_2 x_5, x_2 x_7, x_3 x_4, x_6 x_7)$.
3. If $n = 8$, then $(x_1 x_2, x_1 x_5, x_1 x_8, x_2 x_3, x_3 x_4, x_4 x_5, x_4 x_8, x_5 x_6, x_6 x_7, x_7 x_8)$.

Using a computer with Nauty [16] and CoCoA [17], we classify four-dimensional case: By Theorem 7, it is enough to search for them up to $n = 13$.

**Theorem 8.** (Four-dimensional case) *Let $G$ be a graph with the vertex set $[n]$. Suppose $G$ has no isolate vertex. Then, $S/I(G)^2$ is four-dimensional Cohen-Macaulay if and only if $I(G)$ is one of the following up to the permutation of variables:*

1. If $n = 8$, then $(x_1 x_5, x_2 x_6, x_3 x_7, x_4 x_8)$.
2. If $n = 9$, then $(x_1 x_5, x_2 x_6, x_3 x_7, x_1 x_8, x_4 x_8, x_4 x_9, x_5 x_9)$.
3. If $n = 10$, then

    (a) $(x_1 x_5, x_2 x_6, x_3 x_7, x_1 x_8, x_4 x_8, x_2 x_9, x_4 x_9, x_5 x_9, x_4 x_{10}, x_5 x_{10}, x_6 x_{10})$.

    (b) $(x_1 x_5, x_2 x_6, x_1 x_7, x_3 x_7, x_3 x_8, x_5 x_8, x_2 x_9, x_4 x_9, x_4 x_{10}, x_6 x_{10})$.

4. If $n = 11$, then

    (a) $(x_1 x_5, x_2 x_6, x_3 x_7, x_1 x_8, x_4 x_8, x_2 x_9, x_4 x_9, x_5 x_9, x_3 x_{10}, x_4 x_{10}, x_5 x_{10}, x_6 x_{10}, x_4 x_{11}, x_5 x_{11}, x_6 x_{11}, x_7 x_{11})$.

    (b) $(x_1 x_5, x_2 x_6, x_1 x_7, x_3 x_7, x_3 x_8, x_5 x_8, x_2 x_9, x_4 x_9, x_1 x_{10}, x_4 x_{10}, x_6 x_{10}, x_4 x_{11}, x_5 x_{11}, x_6 x_{11}, x_7 x_{11})$.

5. If $n = 12$, then

    $(x_1 x_5, x_2 x_6, x_1 x_7, x_3 x_7, x_2 x_8, x_4 x_8, x_2 x_9, x_3 x_9, x_5 x_9, x_1 x_{10}, x_4 x_{10}, x_6 x_{10}, x_4 x_{11}, x_5 x_{11}, x_6 x_{11},$
    $x_7 x_{11}, x_3 x_{12}, x_5 x_{12}, x_6 x_{12}, x_8 x_{12})$.

6. If $n = 13$, then

    $(x_1 x_5, x_2 x_6, x_1 x_7, x_3 x_7, x_2 x_8, x_4 x_8, x_2 x_9, x_3 x_9, x_5 x_9, x_1 x_{10}, x_3 x_{10}, x_4 x_{10}, x_6 x_{10}, x_3 x_{11}, x_5 x_{11}, x_6 x_{11},$
    $x_8 x_{11}, x_2 x_{12}, x_4 x_{12}, x_5 x_{12}, x_7 x_{12}, x_4 x_{13}, x_6 x_{13}, x_7 x_{13}, x_9 x_{13})$.

See [18] for the concrete algorithm we used. By Theorem 6 in this case, the Cohen-Macaulay property is equivalent to the $(S_2)$ property, which is independent of the base field $K$.

## 5. Example

In this section, we give an example of a Gorenstein squarefree monomial ideal $I$ such that $S/I^2$ satisfies the Serre condition $(S_2)$, but it is not Cohen-Macaulay.

The Cohen-Macaulay property of $I_\Delta^2$ implies the "Gorenstein" property of $I_\Delta$. More precisely:

**Theorem 9** ([7]). *Let $\Delta$ be a simplicial complex on $[n]$. Suppose that $S/I_\Delta^2$ is Cohen-Macaulay over any field $K$. Then, $\Delta$ is Gorenstein for any field $K$.*

In [7], the authors asked the following question:

**Question.** *Let $\Delta$ be a simplicial complex on $[n]$. Let $S = K[x_1, \ldots, x_n]$ be a polynomial ring for a fixed field $K$. Suppose $\Delta$ satisfies the following conditions:*

1. $\Delta$ is Gorenstein.
2. $S/I_\Delta^2$ satisfies the Serre condition $(S_2)$.

*Then, is it true that $S/I_\Delta^2$ is Cohen-Macaulay?*

Using a list in [19] and CoCoA, we have the following counter-example:

**Example 1.** *Let $K$ be a field of characteristic zero. Set:*

$$I_\Delta = (x_1x_{10}, x_3x_9, x_2x_9, x_7x_8, x_2x_8, x_4x_7, x_5x_6, x_3x_6, x_4x_5, x_6x_8x_{10}, x_2x_5x_{10}, x_1x_4x_9, x_1x_3x_7).$$

*Then, the following conditions hold:*

1. $\Delta$ is Gorenstein.
2. $S/I_\Delta^2$ satisfies the Serre condition $(S_2)$.
3. $S/I_\Delta^2$ is not Cohen-Macaulay.

We explain how to find the example. The manifold page of Lutz [19] gives a classification of all triangulations $\Delta$ of the three-sphere with 10 vertices, which shows that there are 247,882 types. Using Theorem 3, we checked the Serre condition $(S_2)$ for them, and there were only nine types such that $S/I_\Delta^2$ satisfies the Serre condition $(S_2)$. Among the nine types, there was only one simplicial complex $\Delta$ such that $S/I_\Delta^2$ is not Cohen-Macaulay, which is the above example. Note that a triangulation $\Delta$ of a sphere is always Gorenstein. See [18] for more information.

**Author Contributions:** These authors contributed equally to this work.

**Funding:** This work was partially supported by a JSPS Grant-in Aid for Scientific Research (C) 18K03244. The research of Do Trong Hoang was supported by the project ICRTM01 2019.02 of the International Center for Research and Postgraduate Training in Mathematics, VAST.

**Conflicts of Interest:** The authors declare no conflict of interest.

## References

1. Crupi, M.; Rinaldo, G.; Terai, N. Cohen-Macaulay edge ideal whose height is half of the number of vertices. *Nagoya Math. J.* **2011**, *201*, 117–131. [CrossRef]
2. Terai, N.; Yoshida, K. A note on Cohen-Macaulayness of Stanley-Reisner rings with Serre's condition $(S_2)$. *Comm. Algebra* **2008**, *36*, 464–477. [CrossRef]

3. Terai, N.; Trung, N.V. Cohen-Macaulayness of large powers of Stanley-Reisner ideals. *Adv. Math.* **2012**, *229*, 711–730. [CrossRef]
4. Minh, N.C.; Trung, N.V. Cohen-Macaulayness of powers of two-dimensional squarefree monomial ideals. *J. Algebra* **2009**, *322*, 4219–4227. [CrossRef]
5. Trung, N.V.; Tuan, T.M. Equality of ordinary and symbolic powers of Stanley-Reisner ideals. *J. Algebra* **2011**, *328*, 77–93. [CrossRef]
6. Hoang, D.T.; Minh, N.C.; Trung, T.N. Cohen-Macaulay graphs with large girth. *J. Algebra Appl.* **2015**, *14*, 1550112. [CrossRef]
7. Rinaldo, G.; Terai, N.; Yoshida, K. On the second powers of Stanley-Reisner ideals. *J. Commut. Algebra* **2011**, *3*, 405–430. [CrossRef]
8. Crupi, M.; Rinaldo, G.; Terai, N.; Yoshida, K. Effective Cowsik-Nori theorem for edge ideals. *Comm. Algebra* **2010**, *38*, 3347–3357. [CrossRef]
9. Bruns, W.; Herzog, J. *Cohen-Macaulay Rings*; Cambridge Univ. Press: Cambridge, UK, 1993.
10. Stanley, R.P. *Combinatorics and Commutative Algebra*, 2nd ed.; Birkhäuser: Boston, MA, USA; Basel, Switzerland; Stuttgart, Germany, 1996.
11. Villarreal, R.H. *Monomial Algebra*, 2nd ed.; CRC Press: Boca Raton, FL, USA; London, UK; New York, NY, USA, 2015.
12. Pournaki, M.R.; Seyed Fakhari, S.A.; Terai, N.; Yassemi, S. Survey article: Simplicial complexes satisfying Serre's condition: a survey with some new results. *J. Commut. Algebra* **2014**, *6*, 455–483. [CrossRef]
13. Rinaldo, G.; Terai, N.; Yoshida, K. Cohen-Macaulayness for symbolic power ideals of edge ideals. *J. Algebra* **2011**, *347*, 1–22. [CrossRef]
14. Simis, A.; Vasconcelos, W.V.; Villarreal, R.H. On the ideal theory of graphs. *J. Algebra* **1994**, *167*, 389–416. [CrossRef]
15. Hoang, D.T.; Trung, T.N. A characterization of triangle-free Gorenstein graphs and Cohen-Macaulayness of second powers of edge ideals. *J. Algebr. Combin.* **2016**, *43*, 325–338. [CrossRef]
16. McKay, B.D. NAUTY: No AUTomorphisms, Yes? Available online: http://cs.anu.edu.au/{~}bdm/nauty/ (accessed on 29 June 2019).
17. Abbott, J.; Bigatti, A.M.; Robbiano, L. CoCoA: A system for doing Computations in Commutative Algebra. Available online: http://cocoa.dima.unige.it (accessed on 29 June 2019).
18. Hoang, D.T.; Rinaldo, G.; Terai, N. Cohen-Macaulay and S2 of Second Power. Available online: http://www.giancarlorinaldo.it/cm-s2-2nd-power.html (accessed on 29 June 2019).
19. Lutz, F.H. The Manifold Page. Available online: http://page.math.tu-berlin.de/$\sim$lutz/stellar/ (accessed on 29 June 2019).

© 2019 by the authors. Licensee MDPI, Basel, Switzerland. This article is an open access article distributed under the terms and conditions of the Creative Commons Attribution (CC BY) license (http://creativecommons.org/licenses/by/4.0/).

*Article*

# Syzygies, Betti Numbers, and Regularity of Cover Ideals of Certain Multipartite Graphs

**A. V. Jayanthan [1],* and Neeraj Kumar [2]**

1 Department of Mathematics, Indian Institute of Technology Madras, Chennai 600036, India
2 Department of Mathematics, Indian Institute of Technology Hyderabad, Kandi, Sangareddy 502285, India; neeraj@iith.ac.in or neeraj.unix@gmail.com
* Correspondence: jayanav@iitm.ac.in

Received: 10 August 2019; Accepted: 13 September 2019; Published: 19 September 2019

**Abstract:** Let $G$ be a finite simple graph on $n$ vertices. Let $J_G \subset K[x_1,\ldots,x_n]$ be the cover ideal of $G$. In this article, we obtain syzygies, Betti numbers, and Castelnuovo–Mumford regularity of $J_G^s$ for all $s \geq 1$ for certain classes of graphs $G$.

**Keywords:** syzygy; Betti number; Castelnuovo-Mumford regularity; bipartite graph; multipartite graph

## 1. Introduction

In this article, we compute the resolution, syzygies, and Betti numbers of powers of certain classes of squarefree monomial ideals. As a by-product, we obtain the expression for the regularity of powers of these classes of ideals. Computation of minimal free resolution and syzygies of ideals and modules over polynomial rings have always attracted researchers in commutative algebra and algebraic geometry. Recently, there has been much interest in studying the homological aspects of squarefree monomial ideals in polynomial rings. As these ideals have strong combinatorial connections, problems in this area have attracted both commutative algebraists and combinatorists. Even in this case, there are only very few cases of ideals for which explicit computation of the resolution, including the complete description of syzygies, is done. The main open problems in this area are to find/construct minimal free resolutions in more cases, and to introduce new ideas and structures ([1] Remark 2.6). There are even less results on the resolution of powers of ideals. If $I$ is generated by a regular sequence, then $I^s$ is a determinantal ideal and hence the minimal free resolution of $I^s$ can be obtained by using the Eagon–Northcott complex [2]. As the maps in this resolution may not be degree preserving, the computation of graded Betti numbers and more generally the computation of the syzygies may not be possible. In the work by the authors of [3], they explicitly compute the graded Betti numbers of $I^s$ if $I$ is a homogeneous complete intersection in a polynomial ring.

Among the resolutions, linear resolutions are possibly the simplest to describe. Let $R = K[x_1,\ldots,x_n]$, where $K$ is a field. If $I$ is the defining ideal of the rational normal curve in $\mathbb{P}^{n-1}$, then Conca proved that $I^s$ has a linear resolution for all $s$ [4]. Herzog, Hibi, and Zheng proved that if $I$ is a monomial ideal generated in degree 2, then $I$ has linear resolution if and only if $I$ has linear quotients if and only if $I^s$ has a linear resolution for all $s \geq 1$ [5]. It was proved by Conca and Herzog [6], that if $I$ is a polymatroidal ideal in $R$, then $I$ has linear quotients. Moreover, they proved that the product of polymatroidal ideals are polymatroidal. Therefore, if $I$ is polymatroidal, then $I^s$ has linear resolution for all $s \geq 1$. If $I$ is a lexsegment ideal, then Ene and Olteanu proved that $I$ has linear resolution if and only if $I$ has linear quotients if and only if $I^s$ has linear resolution for all $s \geq 1$ if and only if $I^s$ has linear quotients for all $s \geq 1$ [7].

In all the above-mentioned results, the authors do not compute the syzygy modules. In general, it is a nontrivial task to compute the syzygy modules, even when the resolution is linear. It was shown by Kodiyalam [8], and independently by Cutkosky, Herzog, and Trung [9], that if $I$ is a homogeneous ideal in a polynomial ring, then there exist non-negative integers $d, e$ and $s_0$ such that $\reg(I^s) = ds + e$ for all $s \geq s_0$, where $\reg(-)$ denote the Castelnuovo–Mumford regularity. Kodiyalam proved that $d \leq \deg(I)$, where $\deg(I)$ denotes the largest degree of a homogeneous minimal generator of $I$. In general, the stability index $s_0$ and the constant term $e$ are hard to compute. There have been discrete attempts in identifying $s_0$ and $e$ for certain classes of ideals.

Recently, there have been a lot of activity in studying the interplay between the combinatorial properties of graphs and the algebraic properties of ideals associated to graphs. For a finite simple graph $G$ on the vertex set $\{x_1, \ldots, x_n\}$, let $J_G \subset R = K[x_1, \ldots, x_n]$, where $K$ is a field, denote the cover ideal of $G$ (see Section 2 for the definition). It may be noted that $J_G = \cap_{\{x_i, x_j\} \in E(G)}(x_i, x_j)$ is the Alexander dual of the edge ideal of $I$ (see Section 2 for the definition). Although the connection between algebraic properties of the edge ideal and combinatorial properties of the graph has been studied extensively, not much is known about the connection between the properties of the cover ideal and the graph. In [10], Seyed Fakhari studied certain homological properties of symbolic powers of cover ideals of very well covered and bipartite graphs. It was shown that if $G$ is a very well covered graph and $J_G$ has a linear resolution, then $J_G^{(s)}$ has a linear resolution for all $s \geq 1$. Furthermore, it was proved that if $G$ is a bipartite graph with $n$ vertices, then for $s \geq 1$,

$$\reg(J_G^s) \leq s \deg(J_G) + \reg(J_G) + 1.$$

Hang and Trung [11] studied unimodular hypergraphs and proved that if $\mathcal{H}$ is a unimodular hypergraph on $n$ vertices and rank $r$ and $J_\mathcal{H}$ is the cover ideal of $\mathcal{H}$, then there exists a non-negative integer $e \leq \dim(R/J_\mathcal{H}) - \deg(J_\mathcal{H}) + 1$, such that

$$\reg J_\mathcal{H}^s = \deg(J_\mathcal{H})s + e$$

for all $s \geq \frac{rn}{2} + 1$. Since bipartite graphs are unimodular, their results hold true in the case of bipartite graphs as well. While the first result gives an upper bound for the constant term, the later result gives the upper bound for both the stability index and the constant term.

In this article, we obtain the complete description of the minimal free resolution, including the syzygies, of $J_G^s$ for some classes of multipartite graphs. The paper is organized as follows. In Section 2, we collect the preliminaries required for the rest of the paper. We study the resolution of powers of cover ideals of certain bipartite graphs in Section 3. If $G$ is a complete bipartite graph, then $J_G$ is a regular sequence and hence the minimal graded free resolution of $J_G^s$ can be obtained from (Theorem 2.1 in the work by the authors of [3]). We then move on to study some classes of bipartite graphs which are not complete. We obtain the resolution, syzygies and Betti numbers of powers of cover ideals of certain bipartite graphs, and as a by-product, we obtain expression for the regularity of powers of these ideals. Section 4 is devoted to the study of resolution and regularity of powers of cover ideals of certain complete multipartite graphs.

The main results of this article are the following: When $G$ is the cycle of length three or the complete graph on 4 vertices, we describe the graded minimal free resolution of $J_G^s$ for all $s \geq 1$. This allow us to compute the Betti numbers, Hilbert series and the regularity of $J_G^s$ for all $s \geq 1$. As a consequence, for cover ideals of complete tripartite and 4-partite graphs, we obtain precise expressions for the Betti numbers and the regularity of $J_G^s$. We conclude our article with a conjecture on the resolution of $J_G^s$ for all $s \geq 1$, where $G$ is a complete multipartite graph.

## 2. Preliminaries

In this section, we set the notation for the rest of the paper. All the graphs that we consider in this article are finite, simple, and without isolated vertices. For a graph $G$, $V(G)$ denotes the set of all vertices of $G$ and $E(G)$ denotes the set of all edges of $G$. A graph $G$ is said to be a "complete multipartite" graph if $V(G)$ can be partitioned into sets $V_1, \ldots, V_k$ for some $k \geq 2$, such that $\{x, y\} \in E(G)$ if and only if $x \in V_i$ and $y \in V_j$ for $i \neq j$. When $k = 2$, the graph is called a "complete bipartite" graph. If $k = 2$ with $|V_1| = m$ and $|V_2| = n$, we denote the corresponding complete bipartite graph by $K_{m,n}$ or by $K_{V_1, V_2}$. If $G$ and $H$ are graphs, then $G \cup H$ denote the graph on the vertex set $V(G) \cup V(H)$ with $E(G \cup H) = E(G) \cup E(H)$. A graph $G$ is called a "bipartite" graph if $V(G) = V_1 \sqcup V_2$ such that $\{x, y\} \in E(G)$ only if $x \in V_1$ and $y \in V_2$. A subset $w = \{x_{i_1}, \ldots, x_{i_r}\}$ of $V(G)$ is said to be a *vertex cover* of $G$ if $w \cap e \neq \emptyset$ for every $e \in E(G)$. A vertex cover is said to be minimal if it is minimal with respect to inclusion.

Let $G$ be a graph with $V(G) = \{x_1, \ldots, x_n\}$. Let $K$ be a field and $R = K[x_1, \ldots, x_n]$. The edge ideal of $G$ is defined to be $I(G) = \langle\{x_i x_j : \{x_i, x_j\} \in E(G)\}\rangle \subset R$ and the cover ideal of $G$ is defined to be $J_G = \langle x_{i_1}, \ldots, x_{i_r} : \{x_{i_1}, \ldots, x_{i_r}\}$ is a (minimal) vertex cover of $G\rangle$ It can also be seen that $J_G$ is the Alexander dual of $I(G)$.

Let $S = R/I$, where $R$ is a polynomial ring over $K$ and $I$ a homogeneous ideal of $R$. For a finitely generated graded $S$-module $M = \oplus M_i$, set

$$t_i^S(M) = \max\{j : \operatorname{Tor}_i^S(M, K)_j \neq 0\},$$

with $t_i^S(M) = -\infty$ if $\operatorname{Tor}_i^S(M, K) = 0$. The Castelnuovo–Mumford regularity, denoted by $\operatorname{reg}_S(M)$, of an $S$-module $M$ is defined to be

$$\operatorname{reg}_S M = \max\{t_i^S(M) - i : i \geq 0\}.$$

## 3. Bipartite Graphs

In this section, we study the regularity of powers of cover ideals of certain bipartite graphs. We begin with a simple observation concerning the vertex covers of a bipartite graph.

**Proposition 1.** *Let $G$ be a bipartite graph on $n + m$ vertices. Then $G$ is a complete bipartite graph if and only if $J_G$ is generated by a regular sequence that has disjoint supports.*

**Proof.** Let $V(G) = X \sqcup Y$ be the partition of the vertex set of $G$ with $X = \{x_1, \ldots, x_n\}$ and $Y = \{y_{n+1}, \ldots, y_{n+m}\}$. First, note that $J_G$ is generated by a regular sequence if and only if for any two minimal vertex covers $w, w'$, $w \cap w' = \emptyset$. If $G = K_{n,m}$, then $J_G = (x_1 \cdots x_n, y_{n+1} \cdots y_{n+m})$ which is a regular sequence. Conversely, suppose $G$ is not a complete bipartite graph. As $G$ is a bipartite graph, note that $\prod_{x_i \in X} x_i, \prod_{y_j \in Y} y_j \in J_G$ are minimal generators of $J_G$. Therefore, there exist $x_{i_0} \in X$ and $y_{i_0} \in Y$ such that $\{x_{i_0}, y_{i_0}\} \notin E(G)$. Then $w = \{x_i, y_j : i \neq i_0 \text{ and } y_j \in N_G(x_{i_0})\}$ is a minimal vertex cover of $G$ that intersects $X$ as well as $Y$ nontrivially. Therefore $J_G$ is not a complete intersection. □

First we discuss the regularity of powers of cover ideals of complete bipartite graphs. As the cover ideal of a complete bipartite graph is a complete intersection, the result is a consequence of (Theorem 2.1 in the work by the authors of [3]).

**Remark 1.** *Let $J = J_{K_{m,n}}$ be the cover ideal of the complete bipartite graph $K_{m,n}$, $m \leq n$. Then $\operatorname{reg}(J^s) = sn + m - 1$ for all $s \geq 1$.*

**Proof.** Consider the ideal $I = (T_1, T_2) \subset R = K[T_1, T_2]$ with $\deg T_1 = m$ and $\deg T_2 = n$. It follows from (Theorem 2.1 in the work by the authors of [3]) that the resolution of $I^s$ is

$$0 \to \bigoplus_{\substack{a_1 + a_2 = s+1 \\ a_i \geq 1}} R(-a_1 m - a_2 n) \to \bigoplus_{a_1 + a_2 = s} R(-a_1 m - a_2 n) \to I^s \to 0. \tag{1}$$

Note that $J = (x_1 \cdots x_m, y_{m+1} \cdots y_{m+n})$. Set $x_1 \cdots x_m = T_1$ and $y_{m+1} \cdots y_{m+n} = T_2$. Then, $J^s$ has the minimal free resolution as in (1). If $m \leq n$, then $\text{reg}(J^s) = sn + m - 1$. □

It follows from Remark 1 that in the case of the complete bipartite graph $K_{m,n}$, the stability index is 1 and the constant term is $\tau - 1$, where $\tau$ is the size of a minimum vertex cover.

If the graph is not a complete bipartite graph, then the cover ideal is no longer a complete intersection. If $G$ is a Cohen–Macaulay bipartite graph, then it was shown by F. Mohammadi and S. Moradi that the vertex cover ideals are weakly polymatroidal. Therefore, they have linear quotients and hence all the powers have linear resolution (Theorem 2.2 in the work by the authors of [12]). It would be quite a challenging task to obtain the syzygies and Betti numbers of powers of cover ideals of all bipartite graphs. Therefore, we restrict our attention to some structured subclasses of bipartite graphs.

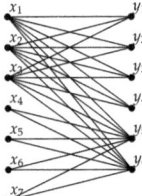

**Figure 1.** $K_{U_1, V} \cup K_{U_2, V_2}$.

**Notation 1.** *For disjoint vertex sets $U$ and $V$, let $K_{U,V}$ denote the complete bipartite graph on $U \sqcup V$. Let $U_1 = \{x_1, \ldots, x_k\}$, $U_2 = \{x_{k+1}, \ldots, x_n\}$, $V_1 = \{y_1, \ldots, y_r\}$, and $V_2 = \{y_{r+1}, \ldots, y_m\}$. Let $U = U_1 \cup U_2$ and $V = V_1 \cup V_2$. In the following, we consider the bipartite graph $G$ on the vertex set $U \sqcup V$ with edges $E(G) = E(K_{U_1,V}) \cup E(K_{U_2,V_2})$. An illustrative figure can be seen in Figure 1 above. Note that although the vertex sets of $K_{U_1,V}$ and $K_{U_2,V_2}$ are not disjoint, the edge sets are. The figure on the left is an example of such a graph with $U_1 = \{x_1, x_2, x_3\}$, $U_2 = \{x_4, x_5, x_6, x_7\}$, $V_1 = \{y_1, y_2, y_3, y_4\}$ and $V_2 = \{y_5, y_6\}$.*

**Theorem 1.** *Let $U = U_1 \sqcup U_2$ and $V = V_1 \sqcup V_2$ be a collection of vertices with $|U| = n$, $|U_i| = n_i$, $|V| = m$, $|V_i| = m_i$ and $1 \leq n_i, m_i$ for $i = 1, 2$. Let $G$ be the bipartite graph $K_{U_1,V} \cup K_{U_2,V_2}$. Let $R = K[x_1, \ldots, x_n, y_1, \ldots, y_m]$. Let $J_G \subset R$ denote the cover ideal of $G$. Then the graded minimal free resolution of $R/J_G$ is of the form*

$$0 \longrightarrow R(-n - m_2) \oplus R(-m - n_1) \longrightarrow R(-(n_1 + m_2)) \oplus R(-m) \oplus R(-n) \longrightarrow R.$$

*In particular,*

$$\text{reg}(J_G) = \max\{n + m_2 - 1, m + n_1 - 1\}$$

**Proof.** It can easily be seen that the cover ideal $J_G$ is generated by $g_1 = x_1 \cdots x_n$, $g_2 = y_1 \cdots y_m$, and $g_3 = x_1 \cdots x_n y_{m_1+1} \cdots y_m$. Set $X_1 = x_1 \cdots x_{n_1}$; $X_2 = x_{n_1+1} \cdots x_n$; $Y_1 = y_1 \cdots y_{m_1}$ and $Y_2 = y_{m_1+1} \cdots y_m$. Then we can write $g_1 = X_1 X_2$, $g_2 = Y_1 Y_2$ and $g_3 = X_1 Y_2$.

Consider the minimal graded free resolution of $R/J_G$ over $R$:

$$\cdots \longrightarrow F \xrightarrow{\partial_2} R(-n) \oplus R(-m) \oplus R(-(n_1 + m_2)) \xrightarrow{\partial_1} R,$$

where $\partial_1(e_1) = g_1, \partial_1(e_2) = g_2$, and $\partial_1(e_3) = g_3$.

Let $ae_1 + be_2 + ce_3 \in \ker \partial_1$. Then $aX_1X_2 + bY_1Y_2 + cX_1Y_2 = 0$. Solving the above equation, it can be seen that,

$$\ker \partial_1 = \text{Span}_R\{Y_2 e_1 - X_2 e_3, X_1 e_2 - Y_1 e_3\}.$$

Also, it is easily verified that these two generators are R-linearly independent. Therefore, $\ker \partial_1 \cong R^2$. Note that $\deg(Y_2 e_1 - X_2 e_3) = n + m_2$ and $\deg(X_1 e_2 - Y_1 e_3) = n_1 + m$. Therefore, we get the minimal free resolution of $R/J_G$ as

$$0 \longrightarrow R(-n-m_2) \oplus R(-m-n_1) \xrightarrow{\partial_2} R(-(n_1+m_2)) \oplus R(-m) \oplus R(-n) \xrightarrow{\partial_1} R,$$

where $\partial_2(a,b) = a(Y_2 e_1 - X_2 e_3) + b(X_1 e_2 - Y_1 e_3)$. The regularity assertion follows immediately from the resolution. □

Our aim is to compute the syzygies and the Betti numbers of $J_G^s$, where $J_G$ is the cover ideal discussed in Theorem 1. For this purpose, we first study the resolution of powers of the ideal $(X_1 X_2, X_1 Y_2, Y_1 Y_2)$ and obtain the resolution and regularity of the cover ideal as a consequence. It is known from the work by the authors of [5] that all the powers of this ideal has a linear resolution. We explicitly compute the syzygies and Betti numbers for the powers of this ideal.

**Theorem 2.** *Let $R = K[X_1, X_2, Y_1, Y_2]$ and $J = (X_1 X_2, X_1 Y_2, Y_1 Y_2)$ be an ideal of R. Then, for $s \geq 2$, the minimal free resolution of $R/J^s$ is of the form*

$$0 \longrightarrow R^{\binom{s}{2}} \longrightarrow R^{2\binom{s+1}{2}} \longrightarrow R^{\binom{s+2}{2}} \longrightarrow R,$$

*and $\mathrm{reg}(J^s) = 2s$.*

**Proof.** Denote the generators of $J$ by $g_1 = X_1 X_2$, $g_2 = X_1 Y_2$, and $g_3 = Y_1 Y_2$. Note that $J$ is the edge ideal of $P_4$, the path graph on the vertices $\{X_1, X_2, Y_1, Y_2\}$. Since $P_4$ is chordal, it follows that $R/J^s$ has a linear resolution for all $s \geq 1$ [5]. Now we compute the Betti number of the $R/J^s$. Write

$$(g_1, g_2, g_3)^s = (g_1^s, g_1^{s-1}(g_2, g_3), g_1^{s-2}(g_2, g_3)^2, \ldots, g_1(g_2, g_3)^{s-1}, (g_2, g_3)^s)$$

where

$$(g_2, g_3)^t = (g_2^t, g_2^{t-1} g_3, g_2^{t-2} g_3^2, \ldots, g_2 g_3^{t-1}, g_3^t).$$

For $i \geq j$, set $M_{i,j} = g_1^{s-i} g_2^{i-j} g_3^j = (X_1 X_2)^{s-i} Y_2^i Y_1^j X_1^{i-j}$. It follows that $\mu(J^s) = \frac{(s+1)(s+2)}{2}$.

Set $\beta_1 = \frac{(s+1)(s+2)}{2}$. Let $\{e_{p,q} \mid 0 \leq p \leq s; 0 \leq q \leq p\}$ denote the standard basis for $R^{\beta_1}$. Let $\partial_1 : R^{\beta_1} \longrightarrow R$ be the map $\partial_1(e_{p,q}) = M_{p,q}$. As $g_i$'s are monomials, the kernel is generated by binomials of the form $m_{i,j} M_{i,j} - m_{k,l} M_{k,l}$, where $m_{p,q}$'s are monomials in $R$. Since the resolution of $R/J^s$ is linear, it is enough to find the linear syzygy relations among the generators of $\ker(\partial_1)$. To find these linear syzygies, we need to find conditions on $i, j, k, l$ such that $\frac{M_{i,j}}{M_{k,l}}$ is equal to $\frac{X_p}{Y_q}$ or $\frac{Y_q}{X_p}$ for some $p, q$. First of all, note that for such linear syzygies, $|i - k|, |j - l| \leq 1$. If $i = k$ and $j = l + 1$, then $\frac{M_{i,j}}{M_{i,j+1}} = \frac{g_3}{g_2} = \frac{Y_1}{X_1}$. We get the same relation if $i = k$ and $j = l - 1$. If $i = k + 1$ and $j = l$, then $\frac{M_{i,j}}{M_{i-1,j}} = \frac{g_2}{g_1} = \frac{Y_2}{X_2}$. As before, $i = k - 1$ yields the same relation. Therefore, the kernel is minimally generated by

$$\{Y_1 e_{i,j} - X_1 e_{i,j+1}, X_2 e_{i,j} - Y_2 e_{i-1,j} \mid 0 \leq j < i \leq s\}.$$

Hence, $\mu(\ker \partial_1) = 2\binom{s+1}{2}$. Write the basis elements of $R^{2\binom{s+1}{2}}$ as

$$\{e_{1,i,p}, e_{2,i,q} \mid 1 \leq i \leq s, 0 \leq p < i \text{ and } 0 \leq q < i\}$$

and define $\partial_2 : R^{2\binom{s+1}{2}} \longrightarrow R^{\beta_1}$ by

$$\begin{aligned}\partial_2(e_{1,i,p}) &= Y_1 e_{i,p} - X_1 e_{i,p+1} \\ \partial_2(e_{2,i,q}) &= X_2 e_{i,q} - Y_2 e_{i-1,q}.\end{aligned}$$

By (Proposition 3.2 in the work by the authors of [13]), $\text{pdim}(R/J^s) = 3$ for all $s \geq 2$. Hence we conclude that the minimal graded free resolution of $R/J^s$ is of the form

$$0 \longrightarrow R^{\beta_3} \longrightarrow R^{2\binom{s+1}{2}} \xrightarrow{\partial_2} R^{\binom{s+2}{2}} \xrightarrow{\partial_1} R.$$

Therefore,

$$\beta_3 - 2\binom{s+1}{2} + \binom{s+2}{2} - 1 = 0,$$

so that $\beta_3 = \binom{s}{2}$. Now we compute the generators of the second syzygy. Again, as the resolution is linear, it is enough to compute linear generators. First, note that

$$B = \{Y_2 e_{1,i,j} - X_2 e_{1,i+1,j} + Y_1 e_{2,i+1,j} - X_1 e_{2,i+1,j+1} \mid 0 \leq j < i < s\} \subseteq \ker \partial_2.$$

It can easily be verified that $B$ is $R$-linearly independent. As $\mu(\ker \partial_2) = \binom{s}{2}$, $B$ generates $\ker \partial_2$. Let $\{E_{i,j} \mid 0 \leq j < i < s\}$ denote the standard basis for $R^{\binom{s}{2}}$. Define $\partial_3 : R^{\binom{s}{2}} \to R^{s(s+1)}$ by

$$\partial_3(E_{i,j}) = Y_2 e_{1,i,j} - X_2 e_{1,i+1,j} + X_1 e_{2,i+1,j} - Y_1 e_{2,i+1,j+1}.$$

Therefore, we get the minimal free resolution of $R/J^s$ as

$$0 \to R^{\binom{s}{2}} \xrightarrow{\partial_3} R^{2\binom{s+1}{2}} \xrightarrow{\partial_2} R^{\binom{s+2}{2}} \xrightarrow{\partial_1} R.$$

As the resolution is linear, $\text{reg } J^s = 2s$. □

As an immediate application of the previous theorem, we obtain resolution and regularity of powers of the cover ideals of graphs discussed in Theorem 1.

**Corollary 1.** *Let $U = U_1 \sqcup U_2$ and $V = V_1 \sqcup V_2$ be a collection of vertices with $|U| = n$, $|U_i| = n_i$, $|V| = m$, $|V_i| = m_i$ and $1 \leq n_i, m_i$ for $i = 1, 2$. Let $G$ be the bipartite graph $K_{U_1,V} \cup K_{U_2,V_2}$. Let $R = K[x_1, \ldots, x_n, y_1, \ldots, y_m]$. Let $J_G \subset R$ denote the cover ideal of $G$. Then the minimal free resolution of $R/J_G^s$ is of the form*

$$0 \longrightarrow R^{\binom{s}{2}} \longrightarrow R^{2\binom{s+1}{2}} \longrightarrow R^{\binom{s+2}{2}} \longrightarrow R.$$

*Moreover,*

$$\text{reg } J_G^s = \max \begin{cases} (s-j)n_1 + (s-i)n_2 + jm_1 + im_2 & \text{for } 0 \leq j \leq i \leq s \\ (s-j)n_1 + (s-i)n_2 + (j+1)m_1 + im_2 - 1 & \text{for } 0 \leq j < i \leq s \\ (s-j)n_1 + (s-i+1)n_2 + jm_1 + im_2 - 1 & \text{for } 0 \leq j < i \leq s \\ (s-j)n_1 + (s-i)n_2 + (j+1)m_1 + (i+1)m_2 - 2 & \text{for } 0 \leq j < i < s \end{cases}.$$

**Proof.** Let $R = K[x_1, \ldots, x_n, y_1, \ldots, y_m]$. Then $J_G = (x_1 \cdots x_n, y_1 \cdots y_m, x_1 \cdots x_{n_1} y_{m_1+1} \cdots y_m)$. Set $X_1 = x_1 \cdots x_{n_1}$, $X_2 = x_{n_1+1} \cdots x_n$, $Y_1 = y_1 \cdots y_{m_1}$ and $Y_2 = y_{m_1+1} \cdots y_m$. Then $J_G = (X_1 X_2, Y_1 Y_2, X_1 Y_2)$. Therefore, it follows from Theorem 1 that $J_G^s$ has the given minimal free resolution.

To compute the regularity, we need to obtain the degrees of the syzygies. Following the notation in the proof of Theorem 1, we can see that

$$\deg e_{i,j} = (s-j)n_1 + (s-i)n_2 + jm_1 + im_2,$$
$$\deg e_{1,i,j} = (s-j)n_1 + (s-i)n_2 + (j+1)m_1 + im_2,$$
$$\deg e_{2,i,j} = (s-j)n_1 + (s-i+1)n_2 + jm_1 + im_2,$$
$$\deg E_{i,j} = (s-j)n_1 + (s-i)n_2 + (j+1)m_1 + (i+1)m_2.$$

Therefore, the assertion on the regularity follows. □

## 3.1. Discussion

It has been proved by Hang and Trung [11] that if $G$ is a bipartite graph on $n$ vertices and $J_G$ is the cover ideal of $G$, then there exists a non-negative integer $e$, such that for $s \geq n+2$, $\text{reg}(J_G^s) = \deg(J_G)s + e$, where $\deg(J_G)$ denote the maximal degree of minimal monomial generators of $J_G$. It follows from Remark 1 that if $G = K_{n,m}$ with $n \geq m$, then $e = m-1$ and the index of stability is 1. If the graph is not a complete bipartite graph, then $e$ does not uniformly represent a combinatorial invariant associated to the graph as can be seen in the computations below. We compute the polynomial $\text{reg}(J_G^s)$ for some classes of bipartite graphs that are considered in Corollary 1. We see that $e$ depends on the relation between the integers $n_1, n_2, m_1$ and $m_2$. Just to illustrate the computation of the polynomial from Corollary 1, we compute $\text{reg}(J_G^s)$ in some cases below:

1. If $m_1 = m_2 = 1$, then it can be seen that for $s \geq 2$,

$$\text{reg}(J_G^s) = \max\{(s-j)n_1 + (s-i+1)n_2 + jm_1 + im_2 - 1 : 0 \leq j < i \leq s\}.$$

Since $(s-j)n_1 + (s-i+1)n_2 + jm_1 + im_2 - 1 = s(n_1 + n_2) + j(1-n_1) + i(1-n_2) + n_2 - 1$ and $n_1 > 1$, $n_2 > 1$, this expression attains maximum when $i$ and $j$ attain minimum, i.e., if $j = 0$ and $i = 1$. Therefore, $\text{reg}(J_G^s) = ns$. Thus, in this case, $e = 0$. It can also be noted that, since $n \geq 2$, it follows from Theorem 1 that $\text{reg}(J_G) = n$. Therefore, in this case, the stability index is also equal to 1.

2. If $n_1 = n_2 = m_1 = m_2 = \ell > 1$, then for $s \geq 2$,

$$\text{reg}(J_G^s) = \max\{(s-j)n_1 + (s-i)n_2 + (j+1)m_1 + (i+1)m_2 - 2 : 0 \leq j < i \leq s\}.$$

Therefore, $\text{reg}(J_G^s) = 2\ell s + (2\ell - 2)$, and therefore $e = 2\ell - 2$. Note that in this case, the stability index is 2.

3. $n_1 \geq m_2 \geq n_2 = m_1$: Note that, in this case, $\deg(J_G) = n_1 + m_2$. We have

$$\text{reg}(J_G^s) = \max\{(s-j)n_1 + (s-i)n_2 + (j+1)m_1 + (i+1)m_2 - 2 : 0 \leq j < i < s\}.$$

Since $(s-j)n_1 + (s-i)n_2 + (j+1)m_1 + (i+1)m_2 - 2 = s(n_1 + n_2) + j(m_1 - n_1) + i(m_2 - n_2) + m_1 + m_2 - 2$. Since $n_1 \geq m_1$ and $m_2 \geq n_2$, the above expression attains the maximum when $i$ attains the maximum and $j$ attains the minimum, i.e., if $i = s-1$ and $j = 0$. Therefore, $\text{reg}(J_G^s) = (n_1 + m_2)s + (n_2 + m_1 - 2)$ and hence $e = n_2 + m_1 - 2$.

4. $n_1 \geq n_2 = m_1 \geq m_2$: In this case, $\deg(J_G) = n$ and

$$\text{reg}(J_G^s) = \max\{(s-j)n_1 + (s-i)n_2 + (j+1)m_1 + (i+1)m_2 - 2 : 0 \leq j < i < s\}.$$

As in the previous case, one can conclude that the maximum is attained when $i = 1$ and $j = 0$. Therefore, $\text{reg}(J_G^s) = ns + (2m_2 - 2)$. Thus $e = 2m_2 - 2$.

5. $n_2 = m_1 \geq n_1 \geq m_2$: As done earlier, one can conclude that $\text{reg}(J_G)^s = ns + (2m_2 - 2)$, and therefore $e = 2m_2 - 2$.

**Remark 2.** *If $G = K_{U_1,V} \cup K_{U_2,V_2} \cup K_{U_3,V_3}$, for some set of vertices $U_i, V_i$, then one can still describe the complete resolution and the regularity of $J_G^s$ using a similar approach. However, the resulting syzygies are not so easy to describe though the generating sets are similar. Therefore, we restrict ourselves to the above discussion.*

## 4. Complete Multipartite Graphs

In this section our goal is to understand the resolution and regularity of the powers of cover ideals of complete $m$-partite graphs. Let $G$ be a complete $m$-partite graph and let $J_G$ be the cover ideal of $G$. Let $R = K[x_1, \ldots, x_m]$. It is known that the cover ideal $J_G$ of complete graph $G = K_m$ is generated by all squarefree monomials $x_1 x_2 \cdots \hat{x}_i \cdots x_m$ of degree $m-1$. Moreover one can also identify this cover ideal with the squarefree Veronese ideal $I = I_{m,m-1}$, and thus it is a polymatroidal ideal [14]. Therefore, by the results of Conca-Herzog [6] and Herzog-Hibi [15], we have

**Remark 3.** *The cover ideal $J_G$ of the complete graph $G = K_m$ has linear quotients and hence has linear resolution. Moreover $J_G^s$ has linear resolution for all $s \geq 1$.*

If $I$ is an ideal of $R$ all of whose powers have linear resolution, then depth $R/I^k$ is a non-increasing function of $k$ and depth $R/I^k$ is constant for all $k \gg 0$, ([14] Proposition 2.1). Further, we have,

**Remark 4.** *(Corollary 3.4 in the work of [14]) Let $R = K[x_1, \cdots, x_m]$ and $J_G$ be the cover ideal of $G = K_m$. Then*
$$\text{depth } R/J_G^s = \max\{0, m-s-1\}.$$

*In particular, depth $R/J_G^s = 0$ for all $s \geq m-1$.*

### 4.1. Complete Tripartite:

We first describe the graded minimal free resolution of $J_{K_3}^s$ for all $s \geq 1$. We also obtain the Hilbert series of the powers.

**Theorem 3.** *Let $R = K[x_1, x_2, x_3]$ and $I = (x_1 x_2, x_1 x_3, x_2 x_3)$. Then, the graded minimal free resolution of $R/I$ is of the form*
$$0 \to R(-3)^2 \to R(-2)^3 \to R.$$

*For $s \geq 2$, the graded minimal free resolution of $R/I^s$ is of the form:*
$$0 \to R(-2s-2)^{\binom{s}{2}} \to R(-2s-1)^{2\binom{s+1}{2}} \to R(-2s)^{\binom{s+2}{2}} \to R$$

*so that $\text{reg}_R(I^s) = 2s$. Moreover, the Hilbert series of $R/I^s$ is given by*
$$H(R/I^s, t) = \frac{1 + 2t + 3t^2 + \cdots + 2st^{2s-1} - \left(\binom{s+2}{2} - 2s - 1\right) t^{2s}}{(1-t)}.$$

**Proof.** It is clear that the resolution of $I$ is as given in the assertion of the theorem. Since $I$ is the cover ideal of $K_3$, $I^s$ has linear minimal free resolution for all $s \geq 1$. Note also that by ([16] Lemma 3.1), depth $R/I^s = 0$ for all $s \geq 2$ so that pdim $R/I^s = 3$ for all $s \geq 2$. Therefore, the minimal free resolution of $R/I^s$ is of the form

$$0 \to R(-2s-2)^{\beta_3} \xrightarrow{\partial_3} R(-2s-1)^{\beta_2} \xrightarrow{\partial_2} R(-2s)^{\beta_1} \xrightarrow{\partial_1} R.$$

Now we describe the syzygies and Betti numbers of $R/I^s$ for $s \geq 2$. Let $g_1 = x_1 x_2, g_2 = x_1 x_3$ and $g_3 = x_2 x_3$. The generators of $I^s$ are of the form $g_1^{\ell_1} g_2^{\ell_2} g_3^{\ell_3}$, where $0 \leq \ell_1, \ell_2, \ell_3 \leq s$ and $\ell_1 + \ell_2 + \ell_3 = s$. Set $f_{\ell_1, \ell_2, \ell_3} = g_1^{\ell_1} g_2^{\ell_2} g_3^{\ell_3} = x_1^{s-\ell_3} x_2^{s-\ell_2} x_3^{s-\ell_1}$. It is easy to see that $\mu(I^s) = \binom{s+2}{2}$, as the cardinality of a minimal generating set of $I^s$ is same as the total number of non-negative integral solution of $\ell_1 + \ell_2 + \ell_3 = s$, which is $\binom{s+2}{2}$.

Let $\{e_{\ell_1,\ell_2,\ell_3} : 0 \leq \ell_1, \ell_2, \ell_3 \leq s; \ell_1 + \ell_2 + \ell_3 = s\}$ denote the standard basis for $R^{\binom{s+2}{2}}$ and consider the map $\partial_1 : R^{\binom{s+2}{2}} \to R$ defined by $\partial_1(e_{\ell_1,\ell_2,\ell_3}) = f_{\ell_1,\ell_2,\ell_3}$. As $f_{\ell_1,\ell_2,\ell_3}$'s are monomials, the kernel is generated by binomials of the form $m_{\ell_1,\ell_2,\ell_3}f_{\ell_1,\ell_2,\ell_3} - m_{t_1,t_2,t_3}f_{t_1,t_2,t_3}$, where $m_{i,j,k}$'s are monomials in $R$. Also, as the minimal free resolution is linear, the kernel is generated in degree 1. Note that $\frac{f_{\ell_1,\ell_2,\ell_3}}{f_{t_1,t_2,t_3}} = x_1^{t_3-\ell_3} x_2^{t_2-\ell_2} x_3^{t_1-\ell_1}$. Therefore, for $\frac{f_{\ell_1,\ell_2,\ell_3}}{f_{t_1,t_2,t_3}}$ to be a linear fraction, $|t_i - \ell_i| \leq 1$ for $i = 1,2,3$.

Let $t_3 = \ell_3$, $t_2 = \ell_2 + 1$ and $t_1 = \ell_1 - 1$. The corresponding linear syzygy relation is

$$x_3 \cdot f_{\ell_1,\ell_2,\ell_3} - x_2 \cdot f_{\ell_1-1,\ell_2+1,\ell_3} = 0, \quad (2)$$

where $1 \leq \ell_1 \leq s$, and $0 \leq \ell_2, \ell_3 \leq s-1$. Note that the number of such relations is equal to the number of integral solution to $(\ell_1 - 1) + \ell_2 + \ell_3 = s$, i.e., $\binom{s+1}{2}$. Similarly, if $t_3 = \ell_3 + 1$, $t_2 = \ell_2 - 1$ and $t_1 = \ell_1$, then we get the corresponding linear syzygy relation as

$$x_2 \cdot f_{\ell_1,\ell_2,\ell_3} - x_1 \cdot f_{\ell_1,\ell_2-1,\ell_3+1} = 0, \quad (3)$$

where $0 \leq \ell_1, \ell_3 \leq s-1$ and $1 \leq \ell_2 \leq s$. Note that the linear syzygy relation obtained by fixing $\ell_2$ and taking $|\ell_i - t_i| = 1$ for $i = 1,2$

$$x_3 \cdot f_{\ell_1,\ell_2,\ell_3} - x_1 \cdot f_{\ell_1-1,\ell_2,\ell_3+1} = 0$$

can be obtained from Equations (2) and (3) by setting the $\ell_i$'s appropriately. Therefore,

$\ker \partial_1 = \langle x_3 \cdot e_{\ell_1,\ell_2,\ell_3} - x_2 \cdot e_{\ell_1-1,\ell_2+1,\ell_3}, \; x_2 \cdot e_{\ell_1-1,\ell_2+1,\ell_3} - x_1 \cdot e_{\ell_1-1,\ell_2,\ell_3+1} : 1 \leq \ell_1 \leq s, \; 0 \leq \ell_2, \ell_3 \leq s-1 \rangle.$

As there are $\binom{s+1}{2}$ minimal generators of type (2) and (3), $\mu(\ker \partial_1) = \beta_2 = 2\binom{s+1}{2}$. Write the standard basis of $R^{2\binom{s+1}{2}}$ as

$$B_2 = \left\{ \begin{array}{l} e_{1,(\ell_2+1,\ell_3),\ell_1-1} \\ e_{2,(\ell_1,\ell_2),\ell_3} \end{array} : 1 \leq \ell_1 \leq s, \; 0 \leq \ell_2, \ell_3 \leq s-1 \right\}$$

and define $\partial_2 : R^{2\binom{s+1}{2}} \to R^{\binom{s+2}{2}}$ by

$$\partial_2(e_{1,(\ell_2+1,\ell_3),\ell_1-1}) = x_2 \cdot e_{\ell_1-1,\ell_2+1,\ell_3} - x_1 \cdot e_{\ell_1-1,\ell_2,\ell_3+1},$$
$$\partial_2(e_{2,(\ell_1,\ell_2),\ell_3}) = x_3 \cdot e_{\ell_1,\ell_2,\ell_3} - x_2 \cdot e_{\ell_1-1,\ell_2+1,\ell_3}.$$

From the equation, $\beta_3 - 2\binom{s+1}{2} + \binom{s+2}{2} - 1 = 0$, it follows that $\beta_3 = \binom{s}{2}$. Now, we describe $\ker \partial_2$. For $2 \leq \ell_1 \leq s$ and $0 \leq \ell_2, \ell_3 \leq s-2$, set

$$E_{\ell_1,\ell_2,\ell_3} = -x_3 \cdot e_{2,(\ell_1-1,\ell_2),\ell_3+1} + x_2 \cdot e_{2,(\ell_1-1,\ell_2+1),\ell_3} + x_2 \cdot e_{1,(\ell_2+2,\ell_3),\ell_1-2} - x_3 \cdot e_{1,(\ell_2+1,\ell_3),\ell_1-1}.$$

Note that

$$B_3 = \{E_{\ell_1,\ell_2,\ell_3} : 2 \leq \ell_1 \leq s, \; 0 \leq \ell_2, \ell_3 \leq s-2\} \subset \ker \partial_2.$$

Let $N = \ker \partial_2$. We claim that $\bar{B}_3 = \{\bar{E}_{\ell_1,\ell_2,\ell_3}\}$ is a basis for $N/\mathfrak{m}N$. It is enough to prove that $\bar{B}_3$ is $R/\mathfrak{m}$-linearly independent. Suppose $\sum_{\ell_1,\ell_2,\ell_3} \alpha_{\ell_1,\ell_2,\ell_3} \bar{E}_{\ell_1,\ell_2,\ell_3} = \bar{0}$. Since $B_3 \subset N_1$ and $\mathfrak{m}N \subset \oplus_{r \geq 2} N_r$, the above equation implies that $\sum \alpha_{\ell_1,\ell_2,\ell_3} E_{\ell_1,\ell_2,\ell_3} = 0$. In the above equation, the coefficient of $e_{2,(i,j),k} = -x_3 \alpha_{i+1,j,i-1} + x_2 \alpha_{i+1,j-1,k}$ and coefficient of $e_{1,(i,j),k} = x_2 \alpha_{k+2,j-2,i} - x_3 \alpha_{k+1,j-1,i}$. As the set $B_2$ is linearly independent in $R^{2\binom{s+1}{2}}$, all of these coefficients have to be zero, i.e., $\alpha_{i,j,k} = 0$ for all $i,j,k$.

This proves our claim and hence $N = \langle B_3 \rangle$. Let $\{H_{\ell_1,\ell_2,\ell_3} : 2 \le \ell_1 \le s, 0 \le \ell_2, \ell_3 \le s-2\}$ denote the standard basis of $R^{\binom{s}{2}}$. Define $\partial_3 : R^{\binom{s}{2}} \to R^{2\binom{s+1}{2}}$ by $\partial_3(H_{\ell_1,\ell_2,\ell_3}) = E_{\ell_1,\ell_2,\ell_3}$. As $\operatorname{pdim}(R/I^s) = 3$, this map is injective and the resolution is complete.

It can also be seen that $\deg e_{\ell_1,\ell_2,\ell_3} = 2s$, $\deg e_{1,(\ell_2+1,\ell_3),\ell_1-1} = \deg e_{2,(\ell_1,\ell_2),\ell_3} = 2s+1$ and $\deg H_{\ell_1,\ell_2,\ell_3} = 2s+2$. Therefore, we get the minimal graded free resolution of $R/I^s$ as

$$0 \to R(-2s-2)^{\binom{s}{2}} \xrightarrow{\partial_3} R(-2s-1)^{2\binom{s+1}{2}} \xrightarrow{\partial_2} R(-2s)^{\binom{s+2}{2}} \xrightarrow{\partial_1} R.$$

Therefore, the Hilbert series of $R/I^s$ is given by

$$\begin{aligned} H(R/I^s, t) &= \frac{1 - \binom{s+2}{2}t^{2s} + 2\binom{s+1}{2}t^{2s+1} - \binom{s}{2}t^{2s+2}}{(1-t)^3} \\ &= (1-t)^{-2} \frac{(1 - \binom{s+2}{2}t^{2s} + 2\binom{s+1}{2}t^{2s+1} - \binom{s}{2}t^{2s+2})}{(1-t)}. \end{aligned}$$

By expanding $(1-t)^{-2}$ in the power series form and multiplying with the numerator, we get the required expression. □

We now proceed to compute the minimal graded free resolution of powers of complete tripartite graphs.

**Notation 2.** Let $G$ denote a complete tripartite graph with $V(G) = V_1 \sqcup V_2 \sqcup V_3$ and $E(G) = \{\{a,b\} : b \in V_i, b \in V_j, i \ne j\}$. Set $V_1 = \{x_1, \ldots, x_\ell\}$, $V_2 = \{y_1, \ldots, y_m\}$ and $V_3 = \{z_1, \ldots, z_n\}$. Let $J_G$ denote the vertex cover ideal of $G$. Let $X = \prod_{i=1}^\ell x_i$, $Y = \prod_{j=1}^m y_j$ and $Z = \prod_{k=1}^n z_i$. It can be seen that $J_G = (XY, XZ, YZ)$.

**Theorem 4.** Let $R = K[x_1, \ldots x_{m_1}, y_1, \ldots, y_{m_2}, z_1, \ldots, z_{m_3}]$. Let $G$ be a complete tripartite graph as in Notation 2. Let $J_G \subset R$ denote the cover ideal of $G$. Then for all $s \ge 2$, the minimal free resolution of $R/J_G^s$ is of the form

$$0 \to R^{\binom{s}{2}} \xrightarrow{\partial_3} R^{2\binom{s+1}{2}} \xrightarrow{\partial_2} R^{\binom{s+2}{2}} \xrightarrow{\partial_1} R.$$

Set $\alpha = (s-\ell_3)m_1 + (s-\ell_2)m_2 + (s-\ell_1)m_3$. Then for all $s \ge 2$,

$$\operatorname{reg}(J_G^s) = \max \begin{cases} \alpha, & \text{for } 0 \le \ell_1, \ell_2, \ell_3 \le s, \\ \alpha + m_3 - 1, & \text{for } 1 \le \ell_1 \le s, 0 \le \ell_2, \ell_3 \le s-1, \\ \alpha + 2m_3 - 2, & \text{for } 2 \le \ell_1 \le s, 0 \le \ell_2, \ell_3 \le s-2. \end{cases}$$

**Proof.** Taking $X_1 = x_1$, $X_2 = x_2$ and $X_3 = x_3$ in Theorem 3, it follows that the minimal free resolution of $S/J_G^s$ is of the given form:

$$0 \to R^{\binom{s}{2}} \xrightarrow{\partial_3} R^{2\binom{s+1}{2}} \xrightarrow{\partial_2} R^{\binom{s+2}{2}} \xrightarrow{\partial_1} R.$$

We now compute the degrees of the generators and hence obtain the regularity. Let $\deg X_1 = m_1$, $\deg X_2 = m_2$, and $\deg X_3 = m_3$. Then, it follows that

$$\deg e_{\ell_1,\ell_2,\ell_3} = \deg\left(X_1^{s-\ell_3} X_2^{s-\ell_3} X_3^{s-\ell_1}\right) = (s-\ell_3)m_1 + (s-\ell_2)m_2 + (s-\ell_1)m_3.$$

We observe that

$$\deg e_{1,(\ell_2+1,\ell_3),\ell_1-1} = \deg e_{2,(\ell_1,\ell_2),\ell_3} = \deg e_{\ell_1,\ell_2,\ell_3} + \deg X_3,$$
$$\deg H_{\ell_1,\ell_2,\ell_3} = \deg e_{\ell_1,\ell_2,\ell_3} + 2\deg X_3.$$

Therefore,

$$\text{reg}(J_G^s) = \max \begin{cases} \deg e_{\ell_1,\ell_2,\ell_3}, & \text{for } 0 \leq \ell_1, \ell_2, \ell_3 \leq s, \\ \deg e_{\ell_1,\ell_2,\ell_3} + \deg(X_3) - 1, & \text{for } 1 \leq \ell_1 \leq s,\ 0 \leq \ell_2, \ell_3 \leq s-1, \\ \deg e_{\ell_1,\ell_2,\ell_3} + 2\deg(X_3) - 2, & \text{for } 2 \leq \ell_1 \leq s,\ 0 \leq \ell_2, \ell_3 \leq s-2, \end{cases}$$

where $\ell_1 + \ell_2 + \ell_3 = s$. Let $\alpha = \deg e_{\ell_1,\ell_2,\ell_3}$. Then the regularity assertion follows. □

Observe that the above expression for $\text{reg}(J_G^s)$ is not really in the form $ds + e$. Given a graph, one can compute $d$ and $e$ by studying the interplay between the cardinality of the partitions. For example, suppose the graph is unmixed, i.e., all the partitions are of same cardinality. Then, $\text{reg}(J_G^s) = 2\ell s + (2\ell - 2)$ for all $s \geq 2$, where $\ell = m_1 = m_2 = m_3$. Note also that the stabilization index in this case is 2. As in Section 3.1, one can derive various expressions for $\text{reg}(J_G^s)$ for different cases as well. Consider the arithmetic progression $m_1 = m + 2r$, $m_2 = m$, and $m_3 = m + r$:

**Corollary 2.** Let $m, r$ be any two positive integers. Let $m_1 = m + 2r$, $m_2 = m$, and $m_3 = m + r$ in Theorem 4. Then for all $s \geq 2$, we have

$$\text{reg}(J_G^s) = s(2m + 3r) + 2m - 2.$$

**Proof.** By Theorem 4, $\alpha = (s - \ell_3)m_1 + (s - \ell_2)m_2 + (s - \ell_1)m_3$. Therefore, we get

$$\alpha = s(3m + 3r) - \ell_3(m + 2r) - \ell_2(m) - \ell_1(m + r).$$

Using Theorem 4, we have for all $s \geq 2$,

$$\text{reg}(J_G^s) = \max \begin{cases} \alpha, & \text{for } 0 \leq \ell_1, \ell_2, \ell_3 \leq s, \\ \alpha + (m + r) - 1, & \text{for } 1 \leq \ell_1 \leq s,\ 0 \leq \ell_2, \ell_3 \leq s-1, \\ \alpha + 2(m + r) - 2, & \text{for } 2 \leq \ell_1 \leq s,\ 0 \leq \ell_2, \ell_3 \leq s-2, \end{cases}$$

where $\ell_1 + \ell_2 + \ell_3 = s$. As regularity $\text{reg}(J_G^s)$ is maximum of all the numbers, we need to maximize the value of $\alpha$. For this to happen, negative terms in $\alpha$ should be minimum. The coefficient of $\ell_3$ is largest among the negative terms in $\alpha$, so $\ell_3$ should be assigned the least value. After $\ell_3$, assign the minimum value to $\ell_1$, and finally take $\ell_2 = s - \ell_1 - \ell_3$. For example, to get the maximum of $\alpha$ when $2 \leq \ell_1 \leq s$, $0 \leq \ell_2, \ell_3 \leq s - 2$, put $\ell_3 = 0$, $\ell_1 = 2$, and $\ell_2 = s - 2$. We get for all $s \geq 2$

$$\text{reg}(J_G^s) = \max \begin{cases} s(2m + 3r), & \text{for } 0 \leq \ell_1, \ell_2, \ell_3, \ell_4 \leq s, \\ s(2m + 3r) + m - 1, & \text{for } 1 \leq \ell_1 \leq s,\ 0 \leq \ell_2, \ell_3, \ell_4 \leq s-1, \\ s(2m + 3r) + 2m - 2, & \text{for } 2 \leq \ell_1 \leq s,\ 0 \leq \ell_2, \ell_3, \ell_4 \leq s-2. \end{cases}$$

For all $m \geq 1$, and for all $s \geq 2$, we get

$$\text{reg}(J_G^s) = s(2m + 3r) + 2m - 2.$$

□

### 4.2. Complete 4-Partite Graphs

We now describe the resolution and regularity of powers of cover ideals of 4-partite graphs. For this purpose, we first study the resolution of powers of cover ideal of the complete graph $K_4$.

**Theorem 5.** Let $R = K[x_1, x_2, x_3, x_4]$ and $I = (x_1x_2x_3, x_1x_2x_4, x_1x_3x_4, x_2x_3x_4)$. Then, for $s \geq 3$, the minimal graded free resolution of $R/I^s$ is of the form

$$0 \to R(-3s - 3)^{\beta_4} \longrightarrow R(-3s - 2)^{\beta_3} \longrightarrow R(-3s - 1)^{\beta_2} \longrightarrow R(-3s)^{\beta_1} \longrightarrow R,$$

where
$$\beta_1 = \binom{s+3}{3}, \ \beta_2 = 3\binom{s+2}{3}, \ \beta_3 = 3\binom{s+1}{3}, \text{ and } \beta_4 = \binom{s}{3}.$$

In particular, $\operatorname{reg}_R(I^s) = 3s$. Moreover the Hilbert series of $R/I^s$ is given by

$$\begin{aligned} H(R/I^s, t) &= \frac{1 + 2t + 3t^2 + 4t^3 + \cdots + 3st^{3s-1} - \left(\binom{s+3}{3} - 3s - 1\right)t^{3s} + \binom{s}{3}t^{3s+1}}{(1-t)^2} \\ &= \frac{\sum_{i=0}^{3s-1}(i+1)t^i - \left(\binom{s+3}{3} - 3s - 1\right)t^{3s} + \binom{s}{3}t^{3s+1}}{(1-t)^2}. \end{aligned}$$

**Proof.** By Remark 3, $I^s$ has linear resolution for all $s \geq 1$. Moreover, by Remark 4, depth $R/I^s = 0$ for all $s \geq 3$, and therefore pdim $R/I^s = 4$ for all $s \geq 3$. Therefore, the minimal graded free resolution of $R/I^s$, for $s \geq 3$, is of the form

$$0 \to R(-3s-3)^{\beta_4} \xrightarrow{\partial_4} R(-3s-2)^{\beta_3} \xrightarrow{\partial_3} R(-3s-1)^{\beta_2} \xrightarrow{\partial_2} R(-3s)^{\beta_1} \xrightarrow{\partial_1} R.$$

Let $g_1 = x_1x_2x_3, g_2 = x_1x_2x_4, g_3 = x_1x_3x_4$ and $g_4 = x_2x_3x_4$. The minimal generators of $I^s$ are of the form $g_1^{\ell_1} g_2^{\ell_2} g_3^{\ell_3} g_4^{\ell_4}$ where $0 \leq \ell_i \leq s$ for every $i$ and $\ell_1 + \ell_2 + \ell_3 + \ell_4 = s$. The number of non-negative integral solution to the linear equation $\ell_1 + \ell_2 + \ell_3 + \ell_4 = s$ is $\binom{s+3}{3}$. Therefore, we have $\mu(J^s) = \beta_1 = \binom{s+3}{3}$. Note that $g_1^{\ell_1} g_2^{\ell_2} g_3^{\ell_3} g_4^{\ell_4} = x_1^{s-\ell_4} x_2^{s-\ell_3} x_3^{s-\ell_2} x_4^{s-\ell_1}$ with $0 \leq \ell_1, \ell_2, \ell_3, \ell_4 \leq s$. Set $f_{\ell_1, \ell_2, \ell_3, \ell_4} = x_1^{s-\ell_4} x_2^{s-\ell_3} x_3^{s-\ell_2} x_4^{s-\ell_1}$. Let

$$\{e_{\ell_1, \ell_2, \ell_3, \ell_4} : 0 \leq \ell_1, \ell_2, \ell_3, \ell_4 \leq s \text{ and } \ell_1 + \ell_2 + \ell_3 + \ell_4 = s\}$$

denote the standard basis of $R^{\binom{s+3}{3}}$ and consider the map $\partial_1 : R^{\binom{s+3}{3}} \to R$ defined by $\partial_1(e_{\ell_1,\ell_2,\ell_3,\ell_4}) = f_{\ell_1,\ell_2,\ell_3,\ell_4}$. Now we find the minimal generators for $\ker \partial_1$.

Let $f_{\ell_1,\ell_2,\ell_3,\ell_4}, f_{t_1,t_2,t_3,t_4}$ be any two minimal monomial generators of $I^s$. It is known that the kernel of $\partial_1$ is generated by binomials of the form $m_{\ell_1,\ell_2,\ell_3,\ell_4} f_{\ell_1,\ell_2,\ell_3,\ell_4} - m_{t_1,t_2,t_3,t_4} f_{t_1,t_2,t_3,t_4}$, where $m_{i,j,k,l}$'s are monomials in $R$. As the syzygy is generated by linear binomials, we need to find conditions on $\ell_1, \ell_2, \ell_3, \ell_4$ such that $\frac{f_{\ell_1,\ell_2,\ell_3,\ell_4}}{f_{t_1,t_2,t_3,t_4}}$ is equal to $\frac{x_i}{x_j}$ for some $i,j$. Observe that

$$\frac{f_{\ell_1,\ell_2,\ell_3,\ell_4}}{f_{t_1,t_2,t_3,t_4}} = x_1^{t_4-\ell_4} x_2^{t_3-\ell_3} x_3^{t_2-\ell_2} x_4^{t_1-\ell_1}.$$

Suppose $t_4 = \ell_4 + 1, t_3 = \ell_3 - 1$ and $t_j = \ell_j$ for $j = 1, 2$. Then we get the linear syzygy relation

$$x_2 f_{\ell_1,\ell_2,\ell_3,\ell_4} - x_1 f_{\ell_1,\ell_2,\ell_3-1,\ell_4+1} = 0. \tag{4}$$

Fixing $t_3$ and $t_4$ with $1 \leq t_3 + t_4 \leq s$, there are as many linear syzygies are there as the number of solutions of $t_1 + t_2 = s - (t_3 + t_4)$. Therefore, for the pair $(t_3, t_4)$, there are $\binom{s+1}{2} + \binom{s}{2} + \cdots + \binom{2}{2} = \binom{s+2}{3}$ number of solutions. Similarly, for each pair $(t_4, t_2), (t_4, t_1), (t_3, t_2), (t_3, t_1)$ and $(t_2, t_1)$, we get $\binom{s+2}{3}$ linear syzygies. Note that the syzygies $x_4 f_{t_1+1, t_2-1, t_3, t_4} - x_3 f_{t_1, t_2, t_3, t_4}$ and $x_3 f_{t_1, t_2, t_3, t_4} - x_2 f_{t_1, t_2-1, t_3+1, t_4}$ give rise to another linear syzygy $x_4 f_{t_1+1, t_2-1, t_3, t_4} - x_2 f_{t_1, t_2-1, t_3+1, t_4}$. The same linear syzygy can also be obtained from a combination of linear syzygies that arise out of the pairs $(t_1, t_4)$ and $(t_3, t_4)$. Therefore, to get a minimal generating set, we only need to consider the linear syzygies corresponding to the pairs $(t_1, t_2), (t_2, t_3)$, and $(t_3, t_4)$. For each such pair, we have $\binom{s+2}{3}$ number of linear syzygies. Therefore, $\beta_2 = 3\binom{s+2}{3}$.

Write the basis elements of $R^{3\binom{s+2}{3}}$ as

$$B_2 = \begin{cases} e_{(1,\ell_1-1,\ell_2),\ell_3+1,\ell_4} \\ e_{(2,\ell_1-1,\ell_4),\ell_2+1,\ell_3} \\ e_{(3,\ell_3,\ell_4),\ell_1,\ell_2} \end{cases} : 1 \leq \ell_1 \leq s,\ 0 \leq \ell_2, \ell_3, \ell_4 \leq s-1 \end{cases}$$

and define $\partial_2 : R^{3\binom{s+1}{2}} \longrightarrow R^{\beta_1}$ by

$$\begin{aligned}
\partial_2(e_{(1,\ell_1-1,\ell_2),\ell_3+1,\ell_4}) &= x_2 e_{\ell_1-1,\ell_2,\ell_3+1,\ell_4} - x_1 e_{\ell_1-1,\ell_2,\ell_3,\ell_4+1}; \\
\partial_2(e_{(2,\ell_1-1,\ell_4),\ell_2+1,\ell_3}) &= x_3 e_{\ell_1-1,\ell_2+1,\ell_3,\ell_4} - x_2 e_{\ell_1-1,\ell_2,\ell_3+1,\ell_4}; \\
\partial_2(e_{(3,\ell_3,\ell_4),\ell_1,\ell_2}) &= x_4 e_{\ell_1,\ell_2,\ell_3,\ell_4} - x_3 e_{\ell_1-1,\ell_2+1,\ell_3,\ell_4}.
\end{aligned}$$

Now, we decipher the Betti numbers $\beta_3$ and $\beta_4$ to complete the resolution. The Hilbert series of $R/I^s$ is

$$H(R/I^s, t) = \frac{(1 - \beta_1 t^{3s} + \beta_2 t^{3s+1} - \beta_3 t^{3s+2} + \beta_4 t^{3s+3})}{(1-t)^4} = \frac{p(t)}{(1-t)^4}.$$

As $\dim R/I^s = 2$, the polynomial $p(t)$ has a factor $(1-t)^2$. Note that $(1-t)^2$ is a monic polynomial of degree 2, therefore we can write $p(t) = (1-t)^2 \cdot q(t)$, where

$$q(t) = \beta_4 t^{3s+1} + (2\beta_4 - \beta_3) t^{3s} + a_{3s-1} t^{3s-1} + \cdots + a_1 t + 1. \tag{5}$$

On the other hand, we also have

$$\begin{aligned}
\frac{p(t)}{(1-t)^2} &= \left( \sum_{n \geq 0} (n+1) t^n \right) \left( 1 - \beta_1 t^{3s} + \beta_2 t^{3s+1} - \beta_3 t^{3s+2} + \beta_4 t^{3s+3} \right) \tag{6} \\
&= \sum_{n=0}^{3s-1} (n+1) t^n + (3s + 1 - \beta_1) t^{3s} + (3s + 2 - 2\beta_1 + \beta_2) t^{3s+1} + \sum_{j \geq 3s+2} a_j t^j
\end{aligned}$$

For the expressions in Equations (5) and (6) to be equal, their respective coefficients should be equal. In particular, we should have that

$$\beta_4 = 3s + 2 - 2\beta_1 + \beta_2 \text{ and } 2\beta_4 - \beta_3 = 3s + 1 - \beta_1.$$

On substituting $\beta_1 = \binom{s+3}{3}$ and $\beta_2 = 3\binom{s+2}{3}$, we get $\beta_4 = \binom{s}{3}$ from the first equation and substituting the value in the second equation, we get $\beta_3 = 3\binom{s+1}{3}$. Now we verify that $a_j = 0$ for all $j \geq 3s + 2$. Note that for $r \geq 2$, the coefficient of $t^{3s+r}$ in Equation (6) is

$$a_{3s+r} = (3s + r + 1) + (r - 2)\beta_4 - (r - 1)\beta_3 + r\beta_2 - (r + 1)\beta_1.$$

As $1 - \beta_1 + \beta_2 - \beta_3 + \beta_4 = 0$, this equation is reduced to $a_{3s+r} = (3s + 3) - \beta_3 + 2\beta_2 - 3\beta_1$. Applying the binomial identify $\binom{n+1}{r+1} = \binom{n}{r+1} + \binom{n}{r}$ repeatedly, we get $a_{3s+r} = 0$ for all $r \geq 2$. Hence the Hilbert series of $R/I^s$ is

$$H(R/I^s, t) = \frac{1 + 2t + 3t^2 + 4t^3 + \cdots + 3st^{3s-1} - \left(\binom{s+3}{3} - 3s - 1\right) t^{3s} + \binom{s}{3} t^{3s+1}}{(1-t)^2}.$$

We now complete the description of the resolution. Write the basis elements of $R^{3\binom{s+1}{3}}$ as

$$B_3 = \begin{cases} E_{1,\ell_1,\ell_2,\ell_3,\ell_4}; \\ E_{2,\ell_1,\ell_2,\ell_3,\ell_4}, \\ E_{3,\ell_1,\ell_2,\ell_3,\ell_4}, \end{cases} : 2 \leq \ell_1 \leq s,\ 0 \leq \ell_2, \ell_3, \ell_4 \leq s - 2 \end{cases}.$$

Note that $|B_3| = 3\binom{s+1}{3}$. Now define the map $\partial_3 : R^{3\binom{s+1}{3}} \longrightarrow R^{3\binom{s+2}{3}}$ by

$$\partial_3(E_{1,\ell_1,\ell_2,\ell_3,\ell_4}) = x_4 e_{(2,\ell_1-1,\ell_4),\ell_2+1,\ell_3} + x_4 e_{(1,\ell_1-1,\ell_2),\ell_3+1,\ell_4} - x_3 e_{(3,\ell_3,\ell_4),\ell_1-1,\ell_2+1}$$
$$- x_3 e_{(2,\ell_1-2,\ell_4),\ell_2+2,\ell_3} - x_3 e_{(1,\ell_1-2,\ell_2+1),\ell_3+1,\ell_4} + x_1 e_{(3,\ell_3,\ell_4+1),\ell_1-1,\ell_2};$$
$$\partial_3(E_{2,\ell_1,\ell_2,\ell_3,\ell_4}) = x_4 e_{(1,\ell_1-1,\ell_2),\ell_3+1,\ell_4} - x_3 e_{(1,\ell_1-2,\ell_2+1),\ell_3+1,\ell_4} - x_2 e_{(3,\ell_3+1,\ell_4),\ell_1-1,\ell_2}$$
$$+ x_1 e_{(3,\ell_3,\ell_4+1),\ell_1-1,\ell_2};$$
$$\partial_3(E_{3,\ell_1,\ell_2,\ell_3,\ell_4}) = x_3 e_{(1,\ell_1-2,\ell_2+1),\ell_3+1,\ell_4} - x_2 e_{(2,\ell_1-2,\ell_4),\ell_2+1,\ell_3+1} - x_2 e_{(1,\ell_1-2,\ell_2),\ell_3+2,\ell_4}$$
$$+ x_1 e_{(2,\ell_1-2,\ell_4+1),\ell_2+1,\ell_3}.$$

We now compute the kernel of $\partial_3$. Consider the set

$$A = \{H_{\ell_1,\ell_2,\ell_3,\ell_4} : 3 \leq \ell_1 \leq s,\ 0 \leq \ell_2, \ell_3, \ell_4 \leq s-3\},$$

where

$$H_{\ell_1,\ell_2,\ell_3,\ell_4} = x_4 E_{3,\ell_1,\ell_2,\ell_3,\ell_4} - x_3 E_{2,\ell_1-1,\ell_2+1,\ell_3,\ell_4} - x_3 E_{3,\ell_1-1,\ell_2+1,\ell_3,\ell_4} + x_2 E_{1,\ell_1-1,\ell_2,\ell_3+1,\ell_4}$$
$$- x_1 E_{1,\ell_1-1,\ell_2,\ell_3,\ell_4+1} + x_1 E_{2,\ell_1-1,\ell_2,\ell_3,\ell_4+1}.$$

It can be verified that $\partial_3(H_{\ell_1,\ell_2,\ell_3,\ell_4}) = 0$, i.e., $A \subseteq \ker \partial_3$. Let $\ell_1' = \ell_1 - 3$, then one has $\ell_1' + \ell_2 + \ell_3 + \ell_4 = s - 3$. The cardinality of $A$ is equal to the total number of non-negative integral solution of this linear equation, which is $\binom{s}{3}$. As in the proof of Theorem 3, it can be seen that $\ker \partial_3 = \langle A \rangle$. Write the basis elements of $R^{\binom{s}{3}}$ as $B_4 = \{G_{\ell_1,\ell_2,\ell_3,\ell_4} : 3 \leq \ell_1 \leq s,\ 0 \leq \ell_2, \ell_3, \ell_4 \leq s-3\}$ and define the map $\partial_4 : R^{\binom{s}{3}} \longrightarrow R^{3\binom{s+1}{3}}$ by

$$\partial_4(G_{\ell_1,\ell_2,\ell_3,\ell_4}) = H_{\ell_1,\ell_2,\ell_3,\ell_4}.$$

This is an injective map, and therefore we get the complete resolution:

$$0 \to R(-3s-3)^{\beta_4} \longrightarrow R(-3s-2)^{\beta_3} \longrightarrow R(-3s-1)^{\beta_2} \longrightarrow R(-3s)^{\beta_1} \longrightarrow R.$$

□

Note that in the above proof, we used $s \geq 3$ only to conclude that $\text{pdim}(R/I^s) = 4$. By Remark 4, $\text{depth}(R/I) = 2$, and therefore $\text{pdim}(R/I) = 2$. Similarly, $\text{depth}(R/I^2) = 1$ and hence $\text{pdim}(R/I^2) = 3$. This forces $\partial_2$ to be injective when $s = 1$ and $\partial_3$ to be injective when $s = 2$. The computations of syzygies in the cases of resolution of $R/I$ and $R/I^2$ remain the same as given in the above proof. Therefore, we get resolutions truncated at $R^{\beta_2}$ in the case of $R/I$ and truncated at $R^{\beta_3}$ in the case of $R/I^2$, with the expressions for $\beta_2$ and $\beta_3$ coinciding with the ones given in the proof. Therefore, we can conclude that in this case, $\text{reg}(I^s) = 3s$ for all $s \geq 1$.

As an immediate consequence, we obtain an expression for the asymptotic regularity of cover ideals of complete 4-partite graphs.

**Theorem 6.** *Let $G$ denote a complete 4-partite graph with $V(G) = \sqcup_{i=1}^{4} V_i$ and $E(G) = \{\{a,b\} : a \in V_i, b \in V_j, i \neq j\}$. Set $V_i = \{x_{i1}, \ldots, x_{im_i}\}$ for $i = 1, \ldots, 4$. Let $J_G \subset R = K[x_{ij} : 1 \leq i \leq 4;\ 1 \leq j \leq m_i]$ denote the cover ideal of $G$. Then the minimal free resolution of $R/J_G^s$ is of the form*

$$0 \to R^{\binom{s}{3}} \longrightarrow R^{3\binom{s+1}{3}} \longrightarrow R^{3\binom{s+2}{3}} \longrightarrow R^{\binom{s+3}{3}} \longrightarrow R.$$

Set $\alpha = (s-\ell_4)m_1 + (s-\ell_3)m_2 + (s-\ell_2)m_3 + (s-\ell_1)m_4$. Furthermore, we have

$$\text{reg}(J_G^s) = \max \begin{cases} \alpha, & \text{for } 0 \leq \ell_1, \ell_2, \ell_3, \ell_4 \leq s, \\ \alpha + m_4 - 1, & \text{for } 1 \leq \ell_1 \leq s,\ 0 \leq \ell_2, \ell_3, \ell_4 \leq s-1, \\ \alpha + 2m_4 - 2, & \text{for } 2 \leq \ell_1 \leq s,\ 0 \leq \ell_2, \ell_3, \ell_4 \leq s-2, \\ \alpha + 3m_4 - 3, & \text{for } 3 \leq \ell_1 \leq s,\ 0 \leq \ell_2, \ell_3, \ell_4 \leq s-3, \end{cases}$$

where $\ell_1 + \ell_2 + \ell_3 + \ell_4 = s$.

**Proof.** Let $X_i = \prod_{j=1}^{m_i} x_{ij}$. Then $J_G = (X_1X_2X_3, X_1X_2X_4, X_1X_3X_4, X_2X_3X_4)$. Then it follows from Theorem 5 that the minimal free resolution of $R/J_G^s$ is of the given form

$$0 \to R^{\binom{s}{3}} \longrightarrow R^{3\binom{s+1}{3}} \longrightarrow R^{3\binom{s+2}{3}} \longrightarrow R^{\binom{s+3}{3}} \longrightarrow R.$$

To compute the regularity of $R/J_G^s$, we first need to find the degree's of the generators of the syzygies. Following the notation of Theorem 5, we have

$$\begin{aligned}
\deg e_{\ell_1,\ell_2,\ell_3,\ell_4} &= \deg\left(X_1^{s-\ell_4} X_2^{s-\ell_3} X_3^{s-\ell_2} X_4^{s-\ell_1}\right) = (s-\ell_4)m_1 + (s-\ell_3)m_2 + (s-\ell_2)m_3 + (s-\ell_1)m_4, \\
\deg e_{(1,\ell_1-1,\ell_2),\ell_3,\ell_4} &= \deg e_{(2,\ell_1-\ell_4),\ell_2+1,\ell_3} = \deg e_{(3,\ell_3,\ell_4),\ell_1,\ell_2} = \deg e_{\ell_1,\ell_2,\ell_3,\ell_4} + \deg(X_4), \\
\deg E_{1,\ell_1,\ell_2,\ell_3,\ell_4} &= \deg E_{2,\ell_1,\ell_2,\ell_3,\ell_4} = \deg E_{3,\ell_1,\ell_2,\ell_3,\ell_4} = \deg e_{\ell_1,\ell_2,\ell_3,\ell_4} + 2\deg(X_4), \\
\deg G_{\ell_1,\ell_2,\ell_3,\ell_4} &= \deg e_{\ell_1,\ell_2,\ell_3,\ell_4} + 3\deg(X_4).
\end{aligned}$$

Therefore, by setting $\alpha = \deg e_{\ell_1,\ell_2,\ell_3,\ell_4}$, we get

$$\text{reg}(J_G^s) = \max \begin{cases} \alpha, & \text{for } 0 \leq \ell_1, \ell_2, \ell_3, \ell_4 \leq s, \\ \alpha + \deg(X_4) - 1, & \text{for } 1 \leq \ell_1 \leq s,\ 0 \leq \ell_2, \ell_3, \ell_4 \leq s-1, \\ \alpha + 2\deg(X_4) - 2, & \text{for } 2 \leq \ell_1 \leq s,\ 0 \leq \ell_2, \ell_3, \ell_4 \leq s-2, \\ \alpha + 3\deg(X_4) - 3, & \text{for } 3 \leq \ell_1 \leq s,\ 0 \leq \ell_2, \ell_3, \ell_4 \leq s-3, \end{cases}$$

where $\ell_1 + \ell_2 + \ell_3 + \ell_4 = s$. □

Here also, we have obtained an expression for $\text{reg}(J_G^s)$ not in the form of a linear polynomial. However, as we have demonstrated in the previous cases, this can always be derived for a given graph. Analyzing the interplay between the cardinalities of the partitions, one can obtain the polynomial expression. Let $m_1 = m_2 = m_3 = m_4 = m$. Then,

$$\begin{aligned}
\alpha &= (s-\ell_4)m_1 + (s-\ell_3)m_2 + (s-\ell_2)m_3 + (s-\ell_1)m_4 \\
&= (4s - (\ell_1 + \ell_2 + \ell_3 + \ell_4))m = 3ms.
\end{aligned}$$

Therefore $\text{reg}(J_G^s) = 3ms + (3m - 3)$ for all $s \geq 3$.

**Corollary 3.** *Let $m, r$ be any two positive integers. Consider the arithmetic progression $m_1 = m$, $m_2 = m + r$, $m_3 = m + 2r$, and $m_4 = m + 3r$ in Theorem 6. Then, for all $s \geq 3$, we have*

$$\text{reg}(J_G^s) = s(3m + 6r) + 3m - 3.$$

**Proof.** We have from Theorem 6, $\alpha = (s-\ell_4)m_1 + (s-\ell_3)m_2 + (s-\ell_2)m_3 + (s-\ell_1)m_4$. On substituting the values of $m_i$'s in $\alpha$, we get

$$\alpha = s(4m + 6r) - m\ell_4 - (m+r)\ell_3 - (m+2r)\ell_2 - (m+3r)\ell_1$$

By Theorem 6, we have for all $s \geq 3$,

$$\operatorname{reg}(J_G^s) = \max \begin{cases} \alpha, & \text{for } 0 \leq \ell_1, \ell_2, \ell_3, \ell_4 \leq s, \\ \alpha + (m+3r) - 1, & \text{for } 1 \leq \ell_1 \leq s,\ 0 \leq \ell_2, \ell_3, \ell_4 \leq s-1, \\ \alpha + 2(m+3r) - 2, & \text{for } 2 \leq \ell_1 \leq s,\ 0 \leq \ell_2, \ell_3, \ell_4 \leq s-2, \\ \alpha + 3(m+3r) - 3, & \text{for } 3 \leq \ell_1 \leq s,\ 0 \leq \ell_2, \ell_3, \ell_4 \leq s-3, \end{cases}$$

where $\ell_1 + \ell_2 + \ell_3 + \ell_4 = s$. To achieve the maximum value of $\alpha$, negative terms in $\alpha$ should be minimum. The coefficient of $\ell_1$ in negative terms in $\alpha$ is largest, so $\ell_1$ should be assigned the minimum value. After assigning the minimum value to $\ell_1$, assign the minimum value to $\ell_2$, and, similarly, the minimum value to $\ell_3$. Then assign $\ell_4 = s - \ell_1 - \ell_2 - \ell_3$. For instance, to get the maximum of $\alpha$ when $1 \leq \ell_1 \leq s$, $0 \leq \ell_2, \ell_3, \ell_4 \leq s-1$, put $\ell_1 = 1, \ell_2 = 0, \ell_3 = 0$, and $\ell_4 = s - 1$. With appropriate substitution, we get for all $s \geq 3$

$$\operatorname{reg}(J_G^s) = \max \begin{cases} s(3m+6r), & \text{for } 0 \leq \ell_1, \ell_2, \ell_3, \ell_4 \leq s, \\ s(3m+6r) + m - 1, & \text{for } 1 \leq \ell_1 \leq s,\ 0 \leq \ell_2, \ell_3, \ell_4 \leq s-1, \\ s(3m+6r) + 2m - 2, & \text{for } 2 \leq \ell_1 \leq s,\ 0 \leq \ell_2, \ell_3, \ell_4 \leq s-2, \\ s(3m+6r) + 3m - 3, & \text{for } 3 \leq \ell_1 \leq s,\ 0 \leq \ell_2, \ell_3, \ell_4 \leq s-3. \end{cases}$$

Clearly for all $m \geq 1$, and for all $s \geq 3$, we get

$$\operatorname{reg}(J_G^s) = s(3m+6r) + 3m - 3.$$

□

### 4.3. Complete m-Partite Graphs

Let $G$ be a complete graph on $m$-vertices. Then the cover ideal $J_G$ of $G$ is generated by $\{x_1 \cdots \hat{x}_i \cdots x_m : 1 \leq i \leq m\}$. It follows from Remark 4 that depth $R/J_G^s = 0$ for all $s \geq m - 1$. Moreover, by Remark 3, we know that $R/J_G^s$ has linear resolution for all $s \geq 1$. Therefore, the minimal graded free resolution of $R/J_G^s$ for all $s \geq m - 1$ is of the form

$$0 \to R(-s(m-1) - m + 1)^{\beta_m} \to \to \cdots \to R(-s(m-1) - 1)^{\beta_2} \to R(-s(m-1))^{\beta_1} \to R.$$

Let $g_1, g_2, \ldots, g_m$ be the minimal generators $J_G$. Then, the elements in $J_G^s$ consist of elements $T_{\ell_1, \ell_2, \ldots, \ell_m} = g_1^{\ell_1} g_2^{\ell_2} \cdots g_m^{\ell_m}$, such that $\ell_1 + \ell_2 + \cdots + \ell_m = s$ and $0 \leq \ell_i \leq s$. Therefore the total number of elements in $J_G^s$ is same as the total number of non-negative integral solution to the linear equation $\ell_1 + \ell_2 + \cdots + \ell_m = s$, which is $\binom{s+m-1}{m-1}$. Thus, $\mu(J_G^s) = \binom{s+m-1}{m-1}$. Therefore, $\beta_1 = \binom{s+m-1}{m-1}$.

Let $\{e_{\ell_1, \ell_2, \ldots, \ell_m} \mid 0 \leq \ell_i \leq s;\text{ and } \ell_1 + \ell_2 + \cdots + \ell_m = s\}$ denote the standard basis for $R^{\beta_1}$. Let $\partial_1 : R^{\beta_1} \longrightarrow R$ be the map $\partial_1(e_{\ell_1, \ell_2, \ldots, \ell_m}) = T_{\ell_1, \ell_2, \ldots, \ell_m}$. As done in the proofs of Theorems 3 and 5, we can see that the first syzygy is given by the relations of the form

$$x_i \cdot T_{\ell_1, \ell_2, \cdots, \ell_{i-1}, \ell_i - 1, \ell_{i+1} + 1, \cdots, \ell_m} - x_{i+1} \cdot T_{\ell_1, \ell_2, \cdots, \ell_{i-1}, \ell_i, \ell_{i+1}, \cdots, \ell_m} = 0$$

for each $1 \leq i \leq m - 1$. Set $\ell_i - 1 = \ell'_i$ and $\ell_{i+1} + 1 = \ell'_{i+1}$. Then it can be seen that, for each $1 \leq i \leq m - 1$, there exist as many such relations as the number of non-negative integer solutions of $\ell_1 + \cdots + \ell'_i + \cdots \ell_m = s - 1$. Therefore, the total number of such linear relations is $(m-1)\binom{s+m-2}{m-1}$. Therefore $\beta_2 = \binom{m-1}{1}\binom{s+m-2}{m-1}$. However it is not very difficult to realize that writing down the higher syzygy relations are quite challenging. Based on Theorems 3 and 5 and some of the experimental results using the computational commutative algebra package Macaulay 2 [17], we propose the following conjecture.

**Conjecture 1.** Let $R = K[x_1, x_2, \ldots, x_m]$ and let $J$ be the cover ideal of the complete graph $K_m$. The minimal graded free resolution of $R/I^s$ for all $s \geq m - 1$ is of the form

$$0 \to R(-s(m-1)-m+1)^{\beta_m} \to \cdots \to R(-s(m-1)-1)^{\beta_2} \to R(-s(m-1))^{\beta_1} \to R,$$

where

$$\beta_i = \binom{m-1}{i-1}\binom{s+m-i}{m-1}.$$

Notice that proving the above conjecture will give the Betti numbers of powers of cover ideals of complete $m$-partite graphs. We conclude our article by proposing an expression for the regularity of powers of the cover ideals of complete $m$-partite graphs:

**Conjecture 2.** Let $G$ denote a complete m-partite graph with $V(G) = \sqcup_{i=1}^{m} V_i$ and $E(G) = \{\{a,b\} : a \in V_i, b \in V_j, i \neq j\}$. Set $V_i = \{x_{i1}, \ldots, x_{in_i}\}$ for $i = 1, \ldots, m$. Let $J_G \subset R = K[x_{ij} : 1 \leq i \leq m; 1 \leq j \leq n_i]$ denote the cover ideal of G. Let $0 \leq \ell_1, \ell_2, \ldots, \ell_m \leq s$ be integers such that $\ell_1 + \ell_2 + \cdots + \ell_m = s$. Set

$$\alpha = s \cdot \left(\sum_{i=1}^{m} n_i\right) - \sum_{i=1}^{m} n_i \ell_{m+1-i}.$$

Then for all $s \geq m - 1$, one has

$$\operatorname{reg}(J_G^s) = \max \begin{cases} \alpha, & \text{for } 0 \leq \ell_1, \ell_2, \cdots, \ell_m \leq s, \\ \alpha + n_m - 1, & \text{for } 1 \leq \ell_1 \leq s, \ 0 \leq \ell_2, \cdots, \ell_m \leq s-1, \\ \alpha + 2(n_m - 1), & \text{for } 2 \leq \ell_1 \leq s, \ 0 \leq \ell_2, \cdots, \ell_m \leq s-2, \\ \quad \vdots \\ \alpha + (m-1)(n_m - 1), & \text{for } m-1 \leq \ell_1 \leq s, \ 0 \leq \ell_2, \cdots, \ell_m \leq s-(m-1). \end{cases}$$

**Author Contributions:** The results here are the outcome of several discussions that we held together. The authors contributed to this work equally.

**Funding:** This research received no external funding.

**Acknowledgments:** Part of the work was done while the second author was visiting Indian Institute of Technology Madras. He would like to thank IIT Madras for their hospitality during the visit. All our computations were done using Macaulay 2. We would like to thank Huy Tài Hà and S. A. Seyed Fakhari for their comments which helped us improve the exposition.

**Conflicts of Interest:** The authors declare no conflict of interest.

### References

1. Peeva, I.; Stillman, M. The minimal free resolution of a Borel ideal. *Expo. Math.* **2008**, *26*, 237–247. [CrossRef]
2. Buchsbaum, D.A.; Eisenbud, D. Generic free resolutions and a family of generically perfect ideals. *Adv. Math.* **1975**, *18*, 245–301. [CrossRef]
3. Guardo, E.; Van Tuyl, A. Powers of complete intersections: graded Betti numbers and applications. *Ill. J. Math.* **2005**, *49*, 265–279. [CrossRef]
4. Conca, A. Hilbert function and resolution of the powers of the ideal of the rational normal curve. *J. Pure Appl. Algebra* **2000**, *152*, 65–74. [CrossRef]
5. Herzog, J.; Hibi, T.; Zheng, X. Monomial ideals whose powers have a linear resolution. *Math. Scand.* **2004**, *95*, 23–32. [CrossRef]
6. Conca, A.; Herzog, J. Castelnuovo-Mumford regularity of products of ideals. *Collect. Math.* **2003**, *54*, 137–152.
7. Ene, V.; Olteanu, A. Powers of lexsegment ideals with linear resolution. *Ill. J. Math.* **2012**, *56*, 533–549. [CrossRef]

8. Kodiyalam, V. Asymptotic behaviour of Castelnuovo-Mumford regularity. *Proc. Amer. Math. Soc.* **2000**, *128*, 407–411. [CrossRef]
9. Cutkosky, S.D.; Herzog, J.; Trung, N. Asymptotic behaviour of the Castelnuovo-Mumford regularity. *Compos. Math.* **1999**, *118*, 243–261. [CrossRef]
10. Seyed Fakhari, S.A. Symbolic powers of cover ideal of very well-covered and bipartite graphs. *arXiv* **2016**, arXiv:1604.0065.
11. Hang, N.T.; Trung, T.N. Regularity of powers of cover ideals of unimodular hypergraphs. *J. Algebra* **2018**, *513*, 159–176. [CrossRef]
12. Mohammadi, F.; Moradi, S. Weakly polymatroidal ideals with applications to vertex cover ideals. *Osaka J. Math.* **2010**, *47*, 627–636.
13. Morey, S. Depths of powers of the edge ideal of a tree. *Comm. Algebra* **2010**, *38*, 4042–4055. [CrossRef]
14. Herzog, J.; Hibi, T. The depth of powers of an ideal. *J. Algebra* **2005**, *291*, 534–550. [CrossRef]
15. Herzog, J.; Hibi, T. *Monomial Ideals, Volume 260 of Graduate Texts in Mathematics*; Springer: London, UK, 2011.
16. Chen, J.; Morey, S.; Sung, A. The stable set of associated primes of the ideal of a graph. *Rocky Mt. J. Math.* **2002**, *32*, 71–89. [CrossRef]
17. Grayson, D.R.; Stillman, M.E. Macaulay2, A Software System for Research in Algebraic Geometry. Available online: http://www.math.uiuc.edu/Macaulay2/ (accessed on 17 February 2018).

© 2019 by the authors. Licensee MDPI, Basel, Switzerland. This article is an open access article distributed under the terms and conditions of the Creative Commons Attribution (CC BY) license (http://creativecommons.org/licenses/by/4.0/).

Article

# Linear Maps in Minimal Free Resolutions of Stanley-Reisner Rings

Lukas Katthän

Goethe-Universität, FB 12–Institut für Mathematik, Postfach 11 19 32, D-60054 Frankfurt am Main, Germany; katthaen@math.uni-frankfurt.de

Received: 17 June 2019; Accepted: 4 July 2019; Published: 6 July 2019

**Abstract:** In this short note we give an elementary description of the linear part of the minimal free resolution of a Stanley-Reisner ring of a simplicial complex $\Delta$. Indeed, the differentials in the linear part are simply a compilation of restriction maps in the simplicial cohomology of induced subcomplexes of $\Delta$. Along the way, we also show that if a monomial ideal has at least one generator of degree 2, then the linear strand of its minimal free resolution can be written using only $\pm 1$ coefficients.

**Keywords:** monomial ideal; Stanley-Reisner ring; linear part

**MSC:** Primary: 05E40; Secondary: 13D02, 13F55

## 1. Introduction

Let $\Bbbk$ be a field and $S = \Bbbk[x_1, \dots, x_n]$ be the polynomial ring over it. Consider a finitely generated graded $S$-module $M$, and its minimal free resolution $\mathbb{F}_\bullet$. The linear part [1] $\mathrm{lin}(\mathbb{F}_\bullet)$ of $\mathbb{F}_\bullet$ has the same modules as $\mathbb{F}_\bullet$, and its differential $d^{\mathrm{lin}}$ is obtained from the differential d of $\mathbb{F}_\bullet$ by deleting all non-linear entries in the matrices representing d in some basis of $\mathbb{F}_\bullet$.

The main result of this short note is an explicit description of $\mathrm{lin}(\mathbb{F}_\bullet)$ in the case where $M = \Bbbk[\Delta]$ is the Stanley-Reisner ring of a simplicial complex $\Delta$. It is well-known that $\mathbb{F}_\bullet$ is multigraded and generated as $S$-module in squarefree multidegrees. By Hochster's formula, it holds that

$$\mathrm{Tor}_i^S(\Bbbk[\Delta], \Bbbk)_U \cong \widetilde{H}^{\#U-i-1}(\Delta_U; \Bbbk),$$

where $U \subseteq \{1, \dots, n\}$ is a squarefree multidegree and $\Delta_U := \{F \in \Delta : F \subseteq U\}$ is the restriction of $\Delta$. Therefore, $\mathrm{lin}(\mathbb{F}_i)$ is isomorphic to the direct sum of modules of the form $\widetilde{H}^{\#U-i-1}(\Delta_U) \otimes_\Bbbk S(-U)$. The differential $d^{\mathrm{lin}}$ turns out to be simply a compilation of all the restriction maps $\widetilde{H}^i(\Delta_U) \to \widetilde{H}^i(\Delta_{U \setminus u}), \omega \mapsto \omega|_{U \setminus u}$, induced by the inclusions $\Delta_{U \setminus u} \subset \Delta_U$.

**Theorem 1.** *Let $\Bbbk[\Delta]$ be the Stanley-Reisner ring of a simplicial complex $\Delta$ and let $\mathbb{F}_\bullet$ denote its minimal free resolution. The linear part $\mathrm{lin}(\mathbb{F}_\bullet)$ of $\mathbb{F}_\bullet$ is isomorphic to the complex with modules*

$$\mathrm{lin}(\mathbb{F}_i) = \bigoplus_{U \subseteq [n]} \widetilde{H}^{\#U-i-1}(\Delta_U) \otimes_\Bbbk S(-U),$$

*and the components of the differential are given by*

$$\begin{array}{rcl}
\widetilde{H}^j(\Delta_U) \otimes_\Bbbk S(-U) & \longrightarrow & \widetilde{H}^j(\Delta_{U \setminus u}) \otimes_\Bbbk S(-U \setminus u) \\
\omega \otimes s & \longmapsto & (-1)^{\alpha(u,U)} \omega|_{U \setminus u} \otimes x_u s
\end{array}$$

This extends the result of Reiner and Welker ([2], Theorem 3.2), which describes the maps in the linear strand of $\mathbb{F}_\bullet$. An alternative description of $\mathrm{lin}(\mathbb{F}_\bullet)$ in terms of the Alexander dual of $\Delta$ was given by Yanagawa ([3], Theorem 4.1).

**Example 1.** *Let $\Delta$ be the simplicial complex with vertex set $\{a,b,c,d,e\}$ and facets $\{a,c,d\}$, $\{b,d,e\}$, $\{c,d,e\}$ and $\{b,c\}$. Its Stanley-Reisner ideal is $I_\Delta = \langle ab, ac, bcd, bce \rangle$. A minimal free resolution $\mathbb{F}_\bullet$ is given by the following complex:*

$$0 \leftarrow S \xleftarrow{\begin{pmatrix} ab & ac & bcd & bce \end{pmatrix}} S^4 \xleftarrow{\begin{pmatrix} -\mathbf{c} & cd & ce & 0 \\ \mathbf{d} & 0 & 0 & 0 \\ 0 & -\mathbf{a} & 0 & e \\ 0 & 0 & -\mathbf{a} & -d \end{pmatrix}} S^4 \xleftarrow{\begin{pmatrix} 0 \\ \mathbf{e} \\ -\mathbf{d} \\ \mathbf{a} \end{pmatrix}} S \leftarrow 0$$

*The linear entries are marked in boldface. We indicate the relevant induced subcomplexes of $\Delta$ in Figure 1. There, the arrows indicate non-zero linear entries in the matrices of $\mathbb{F}_\bullet$. They correspond to non-zero restriction maps in the zero- or one-dimensional cohomology.*

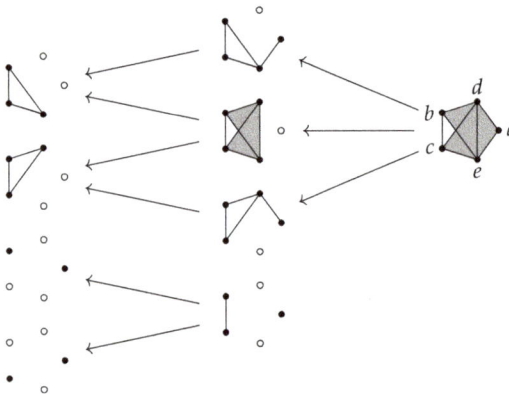

**Figure 1.** The induced subcomplexes of $\Delta$ from Example 1. The arrows indicate non-zero linear coefficients.

As a special case of Theorem 1, we obtain a very simple and explicit description of the 1-linear strand of $\mathbb{F}_\bullet$ (this is the strand containing the quadratic generators of $I_\Delta$). In particular, we show that the maps in the 1-linear strand can always be written using only $\pm 1$ coefficients, see Corollary 1. This extends and simplifies the results of Horwitz [4] and Chen [5], who constructed the minimal free resolution of $I_\Delta$ under the assumption that $I_\Delta$ is generated by quadrics and has a linear resolution.

This article is structured as follows. In Section 2 we set up notational conventions and recall various preliminaries. In the subsequent Section 3 we prove our main result. In the last section, we ask several open questions and pose a conjecture.

## 2. Notation and Preliminaries

For $n \in \mathbb{N}$ we write $[n] := \{1, \ldots, n\}$. To simplify the notation, we set $U \setminus u := U \setminus \{u\}$ and $U \cup u := U \cup \{u\}$ for $U \subseteq [n]$ and $u \in [n]$.

Throughout the paper let $\Bbbk$ denote a fixed field and $S = \Bbbk[x_1, \ldots, x_n]$. Further, we write

$$\mathfrak{m} := \langle x_1, \ldots, x_n \rangle$$

for the unique maximal graded ideal in $S$. We only consider the fine $\mathbb{Z}^n$-grading on $S$, i.e., the degree of $x_i$ is the $i$-th unit vector in $\mathbb{Z}^n$. Squarefree multidegrees are identified with subsets of $[n]$. In particular, for $U \subseteq [n]$, we write $S(-U)$ for the free cyclic $S$-module whose generator is in degree $U$.

*2.1. The Linear Part*

Let $M$ be a finitely generated graded $S$-module. We consider its minimal free resolution

$$\mathbb{F}_\bullet : 0 \longleftarrow M \longleftarrow \mathbb{F}_0 \xleftarrow{d_1} \cdots \xleftarrow{d_n} \mathbb{F}_n \longleftarrow 0.$$

There is a natural filtration on $\mathbb{F}_\bullet$, which is given by

$$\mathcal{F}^j(\mathbb{F}_i) := \mathfrak{m}^{j-i}\mathbb{F}_i.$$

The associated graded complex $\mathrm{lin}(\mathbb{F}_\bullet)$ is called the linear part of $\mathbb{F}_\bullet$. It was introduced in [1], but see also ([6], Chapter 5). Note that $\mathrm{lin}(\mathbb{F}_i) \cong \mathbb{F}_i$ as $S$-modules, but the differentials on the complexes are different. Indeed, $\mathrm{lin}(\mathbb{F}_\bullet)$ can be constructed alternatively by choosing a basis for $\mathbb{F}_\bullet$, representing its differential in this basis by matrices, and deleting all non-linear entries, that is, entries in $\mathfrak{m}^2$.

*2.2. Simplicial Chains and Cochains*

Let $\Delta$ be a simplicial complex with vertex set $[n]$. For the convenience of the reader, we recall the definitions of the chain and cochain complexes of $\Delta$. For keeping track of the signs, we use the notation

$$\alpha(A, B) := \#\{(a, b) \in A \times B : a > b\}$$

for subsets $A, B \subseteq [n]$. We further set $\alpha(a, B) := \alpha(\{a\}, B)$. The (augmented oriented) chain complex of $\Delta$ is the complex of $\Bbbk$-vector spaces $\widetilde{C}_\bullet(\Delta)$, where $\widetilde{C}_d(\Delta)$ is the $\Bbbk$-vector space spanned by the $d$-faces of $\Delta$, and the differential is given by

$$\partial(F) = \sum_{i \in F} (-1)^{\alpha(i,F)} F \setminus i.$$

Here, we consider the empty set as the unique face of dimension $-1$. Note that the definition of $\alpha(i, F)$ depends on the ordering of $[n]$. The *(augmented oriented) cochain complex* of $\Delta$ is the dual complex $\widetilde{C}^\bullet(\Delta) := \hom_\Bbbk(\widetilde{C}_\bullet(\Delta), \Bbbk)$. We write $F^* \in \widetilde{C}^d(\Delta)$ for the basis element dual to a $d$-face $F \in \Delta$. In this basis, the differential on $\widetilde{C}^\bullet(\Delta)$ can be written as

$$\partial(F^*) = \sum_{i \in [n] \setminus F} (-1)^{\alpha(i,F)} (F \cup i)^*.$$

Here, we adopt the convention that $(F \cup i)^* = 0$ if $F \cup i \notin \Delta$. The (reduced) simplicial cohomology of $\Delta$ is $\widetilde{H}^*(\Delta) := \widetilde{H}^*(\Delta; \Bbbk) := H^*(\widetilde{C}^\bullet(\Delta))$.

For a subcomplex $\Gamma \subseteq \Delta$, there is a restriction map $\widetilde{C}^\bullet(\Delta) \to \widetilde{C}^\bullet(\Gamma)$. If $\omega \in \widetilde{C}^\bullet(\Delta)$ is a cochain and $U \subseteq [n]$, then we write $\omega|_U$ for the restriction of $\omega$ to $\Delta_U$.

## 3. Proof of the Main Result

Let $\Delta$ be a simplicial complex with vertex set $[n]$. Recall that the Stanley-Reisner ideal of $\Delta$ is defined as $I_\Delta := \langle x^U : U \subseteq [n], U \notin \Delta \rangle$, where $x^U := \prod_{i \in U} x_i$. Further, the Stanley-Reisner ring is $\Bbbk[\Delta] := S/I_\Delta$. Every squarefree monomial ideal arises as the Stanley-Reisner ideal of some simplicial complex ([7], Theorem 1.7).

We are going to need an explicit version of Hochster's formula. It is of course well known, but we give the details for the convenience of the reader. Let $V = \mathrm{span}_\Bbbk\{e_1, \ldots, e_n\}$ be an $n$-dimensional

$\Bbbk$-vector space and let $\Lambda^\bullet V$ denote the exterior algebra over it. For $F = \{i_1, \ldots, i_r\} \subseteq [n]$ with $i_1 < \cdots < i_r$, we set $\mathbf{e}_F := e_{i_1} \wedge \cdots \wedge e_{i_r}$. Then $\Bbbk[\Delta] \otimes_{\Bbbk} \Lambda^\bullet V$ is the Koszul complex of $\Bbbk[\Delta]$.

**Proposition 1** ([8])**.** *For each squarefree multidegree $U \subseteq [n]$, there is an isomorphism of complexes $(\Bbbk[\Delta] \otimes_{\Bbbk} \Lambda^\bullet V)_U \longrightarrow \widetilde{C}^{\#U-1-\bullet}(\Delta_U)$, given by $x^F \otimes \mathbf{e}_{U\setminus F} \mapsto (-1)^{\alpha(F,U)} F^*$.*

**Proof.** It suffices to show that the following diagram commutes:

$$\begin{array}{ccc}
x^F \otimes \mathbf{e}_{U\setminus F} & \longrightarrow & \sum_{i \in U\setminus F} (-1)^{\alpha(i,U\setminus F)} x^F x_i \otimes \mathbf{e}_{U\setminus (F\cup i)} \\
\downarrow & & \downarrow \\
(-1)^{\alpha(F,U)} F^* & \longrightarrow & (-1)^{\alpha(F,U)} \sum_{i \in U\setminus F} (-1)^{\alpha(i,F)} (F\cup i)^*
\end{array}$$

We only need to show that $\alpha(F, U) + \alpha(i, F) \equiv \alpha(i, U \setminus F) + \alpha(F \cup i, U)$ modulo 2. This follows from the following computation:

$$\alpha(F \cup i, U) - \alpha(F, U) = \alpha(i, U) = \alpha(i, F) + \alpha(i, U \setminus F) \qquad \square$$

Now we turn to the proof of Theorem 1, which we restate for convenience.

**Theorem 2.** *Let $\Bbbk[\Delta]$ be the Stanley-Reisner ring of a simplicial complex $\Delta$ and let $\mathbb{F}_\bullet$ denote its minimal free resolution. The linear part $\mathrm{lin}(\mathbb{F}_\bullet)$ of $\mathbb{F}_\bullet$ is isomorphic to the complex with modules*

$$\mathrm{lin}(\mathbb{F}_i) = \bigoplus_{U \subseteq [n]} \widetilde{H}^{\#U-i-1}(\Delta_U) \otimes_{\Bbbk} S(-U),$$

*and the components of the differential are given by*

$$\widetilde{H}^j(\Delta_U) \otimes_{\Bbbk} S(-U) \longrightarrow \widetilde{H}^j(\Delta_{U\setminus u}) \otimes_{\Bbbk} S(-U \setminus u)$$
$$\omega \otimes s \longmapsto (-1)^{\alpha(u,U)} \omega|_{U\setminus u} \otimes x_u s$$

**Proof of Theorem 1.** We follow the arguments of the proof of ([3], Theorem 4.1). Following [6] and ([1], pp. 107–109), we consider the double complex $(\mathcal{L}_{\bullet,\bullet}, \partial, \partial')$, whose modules are given by $\mathcal{L}_{a,b} := \Bbbk[\Delta] \otimes_{\Bbbk} \Lambda^a V \otimes_{\Bbbk} S_b$ and the differentials are:

$$\partial(s_1 \otimes \mathbf{e}_F \otimes s_2) := \sum_{i \in F} (-1)^{\alpha(i,F)} s_1 x_i \otimes \mathbf{e}_{F\setminus i} \otimes s_2$$

$$\partial'(s_1 \otimes \mathbf{e}_F \otimes s_2) := \sum_{i \in F} (-1)^{\alpha(i,F)} s_1 \otimes \mathbf{e}_{F\setminus i} \otimes x_i s_2$$

It is not difficult to see that the homology of $(\mathcal{L}_{\bullet,\bullet}, \partial)$ is isomorphic to $\mathrm{Tor}^S_\bullet(\Bbbk[\Delta], \Bbbk) \otimes_{\Bbbk} S$. By ([6], Theorem 5.1), the linear part of the minimal free resolution is induced by $\partial'$.

Consider that the sub-double complex $\mathcal{L}'_{a,b} := \bigoplus_{\sigma \in [n]} (\Bbbk[\Delta] \otimes_{\Bbbk} \Lambda^a V \otimes_{\Bbbk} S_b)_\sigma$ of $\mathcal{L}_{\bullet,\bullet}$. As $\mathrm{Tor}^S_\bullet(\Bbbk[\Delta], \Bbbk)$ is non-zero in squarefree degrees only ([7], Corollary 1.40), both $\mathcal{L}'_{\bullet,\bullet}$ and $\mathcal{L}_{\bullet,\bullet}$ have the same homology with respect to $\partial$.

By Proposition 1, $(\mathcal{L}'_{\bullet,\bullet}, \partial)$ is isomorphic to $\bigoplus_{U \subseteq [n]} \widetilde{C}^{\#U-1-\bullet}(\Delta_U) \otimes_{\Bbbk} S(-U)$, where $\partial'$ translates to the map:

$$\widetilde{C}^j(\Delta_U) \otimes_{\Bbbk} S(-U) \longrightarrow \bigoplus_{u \in U} \widetilde{C}^j(\Delta_{U\setminus u}) \otimes_{\Bbbk} S(-U \setminus u)$$
$$F^* \otimes s \longmapsto \sum_{u \in U} (-1)^{\alpha(u,F)} F^*|_{U\setminus u} \otimes x_u s$$

Now the claim follows by taking homology with respect to $\partial$ and applying ([6], Theorem 5.1). □

A particularly simple case of Theorem 1 is the following. See Conjecture 1 for a conjectural improvement of this result.

**Corollary 1.** *Let $I \subseteq S$ be a monomial ideal and let $\mathbb{F}_\bullet$ be its minimal free resolution. Then one can choose a basis of $\mathbb{F}_\bullet$ such that the maps in its 2-linear strand have only coefficients in $\{-1, 0, 1\}$.*

**Proof.** We may assume that $I$ is squarefree by replacing it with its polarization ([7], p. 44). So it is the Stanley-Reisner ideal of some simplicial complex $\Delta$. By Theorem 1, maps in the 2-linear strand of its minimal free resolution are induced by the restriction maps $\widetilde{H}^0(\Delta_U) \to \widetilde{H}^0(\Delta_{U \setminus u})$ for each $u \in U$.

For each subset $U \subseteq [n]$ we choose a distinguished connected component $C_{U,0}$ of $\Delta_U$. For each other connected component $C_{U,i}$ of it, let $e_{U,i} : U \to \mathbb{k}$ the function which is 1 on the vertices of $C_{U,i}$ and 0 on the others. It is clear that the set $\{e_{U,i} : i > 0\}$ forms a basis of $\widetilde{H}^0(\Delta_U)$.

We claim that in this basis, the differential has coefficients $\pm 1$. For $i > 0$ there are the following cases:

1. $C_{U,i} = C_{U \setminus u, j}$ for some $j > 0$,
2. $C_{U,i} = C_{U \setminus u, 0}$,
3. $C_{U,i}$ splits into several connected components $C_{U \setminus u, j_1}, \ldots, C_{U \setminus u, j_r}$ of $\Delta_{U \setminus u}$ with $j_1, \ldots, j_r > 0$,
4. same as (3), with $j_1 = 0$,
5. $C_{U,i}$ is the isolated vertex $u$.

In each case, it is easy to see that $e_{U,i}$ is mapped to a linear combination of the $e_{U \setminus u, j}$ with coefficients in $\{-1, 0, 1\}$. □

## 4. Questions and Open Problems

### 4.1. Affine Monoid Algebras

Recall that a (positive) affine monoid $Q \subseteq \mathbb{N}^n$ is a finitely generated submonoid of $\mathbb{N}^n$. The monoid algebras $\mathbb{k}[Q]$ of affine monoids form a well-studied class of algebras. We refer the reader to [7] or ([9], Chapter 6) for more information on these rings. Each positive affine monoid has a unique minimal generating set, which is called its Hilbert basis. It yields a set of generators for $\mathbb{k}[Q]$ and thus a surjection $S \to \mathbb{k}[Q]$ from a polynomial ring $S$. Moreover, $\mathbb{k}[Q]$ carries a natural $\mathbb{N}^n$-multigrading. There is a combinatorial interpretation of the multigraded Betti numbers of $\mathbb{k}[Q]$, namely $\mathrm{Tor}_i^S(\mathbb{k}[Q], \mathbb{k})_\mathbf{a} \cong \widetilde{H}_i(\Delta_\mathbf{a})$, for a certain simplicial complex $\Delta_\mathbf{a}$, see ([7], Theorem 9.2).

**Question 1.** *Is there a topological interpretation of the linear part of the minimal free resolution of $\mathbb{k}[Q]$ over $S$?*

In this situation, a description along the lines of Theorem 1 would require a map $\widetilde{H}_i(\Delta_\mathbf{a}) \to \widetilde{H}_{i-1}(\Delta_{\mathbf{a}-\mathbf{b}})$, where $\mathbf{b}$ is an element of the Hilbert basis such that $\mathbf{a} - \mathbf{b} \in Q$. Here, $\Delta_{\mathbf{a}-\mathbf{b}}$ is a subcomplex of $\Delta_\mathbf{a}$, but in general it is neither a restriction nor a link.

### 4.2. Approximations of Resolutions

Let $I_\Delta \subseteq S$ be the Stanley-Reisner ideal of some simplicial complex $\Delta$ and let $\mathbb{F}_\bullet$ denote the minimal free resolution of $I_\Delta$. Hochster's formula can be interpreted as giving a description of the complex $\mathbb{F}_\bullet / \mathfrak{m} \mathbb{F}_\bullet$ (with trivial differential). Our Theorem 1 extends this by (essentially) describing $\mathbb{F}_\bullet / \mathfrak{m}^2 \mathbb{F}_\bullet$. These results can be considered as successive approximations of $\mathbb{F}_\bullet$, so the following question seems natural:

**Question 2.** *Is there a combinatorial or topological description of $\mathbb{F}_\bullet / \mathfrak{m}^3 \mathbb{F}_\bullet$?*

This seems to be substantially more difficult than describing $\mathbb{F}_\bullet/\mathfrak{m}^2\mathbb{F}_\bullet$. One reason for this is the following. Even though a minimal free resolution is unique up to isomorphism, if one wants to write it down explicitly one needs to choose an $S$-basis for $\mathbb{F}_\bullet$. This choice can be done in two steps. First choose a $\Bbbk$-basis for $\mathbb{F}_\bullet/\mathfrak{m}\mathbb{F}_\bullet = \operatorname{Tor}_*(S/I_\Delta, \Bbbk)$, and then choose a lifting of these elements to $\mathbb{F}_\bullet$ (any such lifting works due to Nakayama's lemma). Hochster's formula is a convenient tool for the first choice. Theorem 1 implies that the differential of $\mathbb{F}_\bullet/\mathfrak{m}^2\mathbb{F}_\bullet$ does not depend on the second choice, but this is no longer true for $\mathbb{F}_\bullet/\mathfrak{m}^3\mathbb{F}_\bullet$.

*4.3. Coefficients in Resolutions*

Let $I \subseteq S$ be a monomial ideal containing no variables, and let $\mathbb{F}_\bullet$ denote it with minimal free resolution. We saw in Corollary 1 that the differential in the 2-linear strand of $\mathbb{F}_\bullet$ can be written using only coefficients $\pm 1$. On the other hand, in ([2], Section 5) Reiner and Welker gave an example where the differential on the 4-linear strand cannot be written using only coefficients $\pm 1$. We believe that their example is optimal in that sense, and hence offer the following conjecture.

**Conjecture 1.** *Let $I \subseteq S$ be a monomial ideal. Then it is possible to choose a basis for its minimal free resolution $\mathbb{F}_\bullet$, such that the differential on the 3-linear strand can be written using only coefficients $\pm 1$.*

Note that the first map in $\mathbb{F}_\bullet$, $d: \mathbb{F}_1 \to \mathbb{F}_0$, can always be written using coefficients from $\{-1, 0, 1\}$. This is easily seen by considering the Taylor resolution. Further, it is not difficult to explicitly give a basis for $\mathbb{F}_2$ such that the differential $d: \mathbb{F}_2 \to \mathbb{F}_1$ has coefficients $\pm 1$.

**Funding:** This research was supported by the Deutsche Forschungsgemeinschaft (DFG), grant KA 4128/2-1.

**Acknowledgments:** The author thanks Vic Reiner and Srikanth Iyengar for inspiring discussions.

**Conflicts of Interest:** The author declares no conflict of interest.

## References

1. Eisenbud, D.; Fløystad, G.; Schreyer, F.O. Sheaf cohomology and free resolutions over exterior algebras. *Trans. Am. Math. Soc.* **2003**, *355*, 4397–4426. [CrossRef]
2. Reiner, V.; Welker, V. Linear syzygies of Stanley-Reisner ideals. *Math. Scand.* **2001**, *89*, 117–132. [CrossRef]
3. Yanagawa, K. Alexander Duality for Stanley–Reisner rings and squarefree $\mathbb{N}^n$-graded modules. *J. Algebra* **2000**, *225*, 630–645. [CrossRef]
4. Horwitz, N. Linear resolutions of quadratic monomial ideals. *J. Algebra* **2007**, *318*, 981–1001. [CrossRef]
5. Chen, R.X. Minimal free resolutions of linear edge ideals. *J. Algebra* **2010**, *324*, 3591–3613. [CrossRef]
6. Herzog, J.; Simis, A.; Vasconcelos, W.V. Approximation complexes of blowing-up rings, II. *J. Algebra* **1983**, *82*, 53–83. [CrossRef]
7. Miller, E.; Sturmfels, B. *Combinatorial Commutative Algebra*; Springer: New York, NY, USA, 2005.
8. Hochster, M. Cohen-Macaulay rings, combinatorics, and simplicial complexes. In *Ring Theory II*; Dekker: New York, NY, USA, 1977; pp. 171–223.
9. Bruns, W.; Herzog, J. *Cohen-Macaulay Rings*; Cambridge University Press: Cambridge, UK, 1998.

© 2019 by the author. Licensee MDPI, Basel, Switzerland. This article is an open access article distributed under the terms and conditions of the Creative Commons Attribution (CC BY) license (http://creativecommons.org/licenses/by/4.0/).

*Article*

# Toric Rings and Ideals of Stable Set Polytopes

**Kazunori Matsuda [1], Hidefumi Ohsugi [2,\*] and Kazuki Shibata [3]**

1  Faculty of Engineering, Kitami Institute of Technology, Kitami, Hokkaido 090-8507, Japan
2  Department of Mathematical Sciences, School of Science and Technology, Kwansei Gakuin University, Sanda, Hyogo 669-1337, Japan
3  Department of Mathematics, College of Science, Rikkyo University, Toshima-ku, Tokyo 171-8501, Japan
\*  Correspondence: ohsugi@kwansei.ac.jp

Received: 6 June 2019; Accepted: 8 July 2019; Published: 10 July 2019

**Abstract:** In the present paper, we study the normality of the toric rings of stable set polytopes, generators of toric ideals of stable set polytopes, and their Gröbner bases via the notion of edge polytopes of finite nonsimple graphs and the results on their toric ideals. In particular, we give a criterion for the normality of the toric ring of the stable set polytope and a graph-theoretical characterization of the set of generators of the toric ideal of the stable set polytope for a graph of stability number two. As an application, we provide an infinite family of stable set polytopes whose toric ideal is generated by quadratic binomials and has no quadratic Gröbner bases.

**Keywords:** toric ideals; Gröbner bases; graphs; stable set polytopes

## 1. Introduction

Let $\mathcal{P} \subset \mathbb{R}^n$ be an integral convex polytope, i.e., a convex polytope whose vertices have integer coordinates, and let $\mathcal{P} \cap \mathbb{Z}^n = \{\mathbf{a}_1, \ldots, \mathbf{a}_m\}$. Let $K[X, X^{-1}, t] = K[x_1, x_1^{-1}, \ldots, x_n, x_n^{-1}, t]$ be the Laurent polynomial ring in $n+1$ variables over a field $K$. Given an integer vector $\mathbf{a} = (a_1, \ldots, a_n) \in \mathbb{Z}^n$, we set $X^{\mathbf{a}} t = x_1^{a_1} \cdots x_n^{a_n} t \in K[X, X^{-1}, t]$. Then, the *toric ring* of $\mathcal{P}$ is the subalgebra $K[\mathcal{P}]$ of $K[X, X^{-1}, t]$ generated by $\{X^{\mathbf{a}_1} t, \ldots, X^{\mathbf{a}_m} t\}$ over $K$. Here, we need the variable $t$ in order to regard $K[\mathcal{P}]$ as a homogeneous algebra by setting each $\deg X^{\mathbf{a}_i} t = 1$. The *toric ideal* $I_{\mathcal{P}}$ of $\mathcal{P}$ is the kernel of a surjective homomorphism $\pi : K[y_1, \ldots, y_m] \to K[\mathcal{P}]$ defined by $\pi(y_i) = X^{\mathbf{a}_i} t$ for $1 \leq i \leq m$. In general, $I_{\mathcal{P}}$ is generated by homogeneous binomials and any reduced Gröbner basis of $I_{\mathcal{P}}$ consists of homogeneous binomials; see [1]. A simplex $\sigma$ is called a *subsimplex* of $\mathcal{P}$ if the set of vertices of $\sigma$ is contained in $\mathcal{P} \cap \mathbb{Z}^n$. A set $\Delta$ of subsimplices of $\mathcal{P}$ is called a *covering* of $\mathcal{P}$ if $\bigcup_{\sigma \in \Delta} \sigma = \mathcal{P}$. A covering $\Delta$ of $\mathcal{P}$ is called a *triangulation* of $\mathcal{P}$ if $\Delta$ is a simplicial complex. A covering (triangulation) $\Delta$ of $\mathcal{P}$ is called *unimodular* if the normalized volume of each maximal simplex in $\Delta$ is equal to 1. The following properties of an integral convex polytope $\mathcal{P}$ have been investigated in many papers on commutative algebra and combinatorics:

(i)  $\mathcal{P}$ is unimodular (any triangulation of $\mathcal{P}$ is unimodular), the initial ideal of $I_{\mathcal{P}}$ is generated by squarefree monomials with respect to any monomial order ([2] Section 4.3);
(ii)  $\mathcal{P}$ is compressed (any "pulling" triangulation is unimodular), the initial ideal of $I_{\mathcal{P}}$ is generated by squarefree monomials with respect to any reverse lexicographic order [3,4];
(iii)  $\mathcal{P}$ has a regular unimodular triangulation, there exists a monomial order such that the initial ideal of $I_{\mathcal{P}}$ is generated by squarefree monomials ([2] Theorem 4.17);
(iv)  $\mathcal{P}$ has a unimodular triangulation ([2] Section 4.2.4);
(v)  $\mathcal{P}$ has a unimodular covering ([2] Section 4.2.4);
(vi)  $\mathcal{P}$ is normal, $K[\mathcal{P}]$ is a normal semigroup ring ([2] Section 4.2.3).

Details for these conditions are explained in ([1] Chapter 8) and ([2] Chapter 4). The hierarchy (i) $\Rightarrow$ (ii) $\Rightarrow$ (iii) $\Rightarrow$ (iv) $\Rightarrow$ (v) is trivial by their definition. The implication (v) $\Rightarrow$ (vi) is explained in ([2] Theorem 4.11). Note that the converse of each of the above implications is false. On the other hand, the following properties of $I_\mathcal{P}$ are studied by many authors:

(a) $I_\mathcal{P}$ has a Gröbner bases consisting of quadratic binomials;
(b) $K[\mathcal{P}]$ is Koszul ([2] Definition 2.20);
(c) $I_\mathcal{P}$ is generated by quadratic binomials.

The hierarchy (a) $\Rightarrow$ (b) $\Rightarrow$ (c) is known, and the converse of each of the two implications is false; see ([5] Examples 2.1 and 2.2) and [6].

The purpose of this paper is to study such properties of toric rings and ideals of stable set polytopes of simple graphs. Let $G$ be a finite simple graph on the vertex set $[n] = \{1, 2, \ldots, n\}$, and let $E(G)$ denote the set of edges of $G$. Given a subset $W \subset [n]$, we associate the $(0, 1)$ vector $\rho(W) = \sum_{j \in W} \mathbf{e}_j \in \mathbb{R}^n$, where $\mathbf{e}_i$ is the $i$th unit vector of $\mathbb{R}^n$. In particular, $\rho(\emptyset) = \mathbf{0} \in \mathbb{R}^n$. A subset $W \subset [n]$ is said to be *stable* (or *independent*) if $\{i, j\} \notin E(G)$ for all $i, j \in W$ with $i \neq j$. In particular, $\emptyset$ and each $\{i\}$ with $i \in [n]$ are stable. Let $S(G)$ be the set of all stable sets of $G$. The *stable set polytope* (*independent set polytope*) $\mathcal{Q}_G \subset \mathbb{R}^n$ of a simple graph $G$ is the convex hull of $\{\rho(W) : W \in S(G)\}$.

**Example 1.** *If $G$ is a complete graph, then $\mathcal{Q}_G$ is a unit simplex, and hence $I_{\mathcal{Q}_G} = \{0\}$ and $K[\mathcal{Q}_G] = K[x_1 t, \ldots, x_n t, t] \simeq K[y_1, \ldots, y_{n+1}]$.*

Stable set polytopes are very important in many areas, such as optimization theory as well as combinatorics and commutative algebra. Below, we present a list of results on the toric ring $K[\mathcal{Q}_G]$ and the toric ideals $I_{\mathcal{Q}_G}$ of the stable set polytope $\mathcal{Q}_G$ of a simple graph $G$.

1. The stable set polytope $\mathcal{Q}_G$ is compressed if and only if $G$ is perfect ([3,4,7]).
2. Let $G$ be a perfect graph. Then, the toric ring $K[\mathcal{Q}_G]$ is Gorenstein if and only if all maximal cliques of $G$ have the same cardinality ([8]).
3. The toric ring $K[\mathcal{Q}_G]$ is strongly Koszul ([2] p. 53) if and only if $G$ is trivially perfect ([9] Theorem 5.1).
4. Let $G(P)$ be a comparability graph of a poset $P$. Then, $\mathcal{Q}_{G(P)}$ is called a *chain polytope* of $P$. It is known that the toric ideals of a chain polytope have a squarefree quadratic initial ideal (see [10] Corollary 3.1). For example, if a graph $G$ is bipartite, then there exists a poset $P$ such that $G = G(P)$.
5. Suppose that a graph $G$ on the vertex set $[n]$ is an almost bipartite graph, i.e., there exists a vertex $v$ such that the induced subgraph of $G$ on the vertex set $[n] \setminus \{v\}$ is bipartite. Then, $I_{\mathcal{Q}_G}$ has a squarefree quadratic initial ideal ([11] Theorem 8.1). For example, any cycle is almost bipartite.
6. Let $G$ be the complement of an even cycle of length $2k$. Then, the maximum degree of a minimal set of binomial generators of $I_{\mathcal{Q}_G}$ is equal to $k$ ([11] Theorem 7.4).

In the present paper, we study the normality of the toric rings of stable set polytopes, generators of toric ideals of stable set polytopes, and their Gröbner bases via the notion of edge polytopes of finite (nonsimple) graphs and the results on their toric ideals. Here, the *edge polytope* $\mathcal{P}_G \subset \mathbb{R}^n$ of a graph $G$ allowing loops and having no multiple edges is the convex hull of

$$\{\mathbf{e}_i + \mathbf{e}_j : \{i, j\} \text{ is an edge of } G\} \cup \{2\mathbf{e}_i : \text{ there is a loop at } i \text{ in } G\}.$$

This paper is organized as follows. In Section 2, fundamental properties of $K[\mathcal{Q}_G]$ and $I_{\mathcal{Q}_G}$ are studied. In particular, the relationship between the stable set polytopes and the edge polytopes are given (Lemma 1). In addition, it is shown that $\mathcal{Q}_G$ is unimodular if and only if the complement of $G$ is bipartite (Proposition 5). We also point out that, by the results in [11], it is easy to see that $I_{\mathcal{Q}_G}$

has a squarefree quadratic initial ideal if $G$ is either a chordal graph or a ring graph (Proposition 2). In Section 3, we discuss the normality of the stable set polytopes. We prove that, for a simple graph $G$ of stability number two, $\mathcal{Q}_G$ is normal if and only if the complement of $G$ satisfies the "odd cycle condition" (Theorem 1). Using this criterion, we construct an infinite family of normal stable set polytopes without regular unimodular triangulations (Theorem 2). For general simple graphs, some necessary conditions for $\mathcal{Q}_G$ to be normal are also given. In Section 4, we study the set of generators and Gröbner bases of toric ideals of stable set polytopes. It is shown that for a simple graph $G$ of stability number two, the set of binomial generators of $I_{\mathcal{Q}_G}$ are described in terms of even closed walks of a graph (Theorem 3). If $\overline{G}$ is bipartite and if $I_{\mathcal{Q}_G}$ is generated by quadratic binomials, then $I_{\mathcal{Q}_G}$ has a quadratic Gröbner basis (Corollary 1). Finally, using the results on normality, generators, and Gröbner bases, we present an infinite family of non-normal stable set polytopes whose toric ideal is generated by quadratic binomials and has no quadratic Gröbner bases (Theorem 4).

## 2. Fundamental Properties of the Stable Set Polytopes

In this section, we give some fundamental properties of $K[\mathcal{Q}_G]$ and $I_{\mathcal{Q}_G}$. In particular, a relation between the stable set polytopes and the edge polytopes is discussed. The *stability number* $\alpha(G)$ of a graph $G$ is the cardinality of the largest stable set. If a simple graph $G$ satisfies $\alpha(G) = 1$, then $G$ is a complete graph in Example 1. Given lattice polytopes $\mathcal{P}_1 \subset \mathbb{R}^m$ and $\mathcal{P}_2 \subset \mathbb{R}^n$, the *product* of $\mathcal{P}_1$ and $\mathcal{P}_2$ is defined by $\mathcal{P}_1 \times \mathcal{P}_2 = \{(\alpha_1, \alpha_2) \in \mathbb{R}^{m+n} : \alpha_1 \in \mathcal{P}_1, \alpha_2 \in \mathcal{P}_2\}$. Then the toric ring $K[\mathcal{P}_1 \times \mathcal{P}_2]$ is called the *Segre product* of $K[\mathcal{P}_1]$ and $K[\mathcal{P}_2]$.

**Example 2.** *Suppose that a simple graph $G$ is not connected. Let $G_1, \ldots, G_s$ be the connected components of $G$. Then, it is easy to see that $\mathcal{Q}_G = \mathcal{Q}_{G_1} \times \cdots \times \mathcal{Q}_{G_s}$, and hence $K[\mathcal{Q}_G]$ is the Segre product of $K[\mathcal{Q}_{G_1}], \ldots, K[\mathcal{Q}_{G_s}]$.*

Thus, it is enough to study stable set polytopes of connected simple graphs $G$ such that $\alpha(G) \geq 2$. We use the notion of toric fiber products to study toric rings of stable set polytopes. This notion is first introduced in [12] as a generalization of the Segre product. Since the definition of toric fiber products is complicated, we give an example.

**Example 3.** *Let $G_1$ and $G_2$ be cycles of length 4, where $E(G_1) = \{\{1,2\},\{2,3\},\{3,4\},\{1,4\}\}$ and $E(G_2) = \{\{3,4\},\{4,5\},\{5,6\},\{3,6\}\}$. Then $K[\mathcal{Q}_{G_1}] = K[t, x_1t, x_2t, x_3t, x_4t, x_1x_3t, x_2x_4t]$ and $K[\mathcal{Q}_{G_2}] = K[t, x_3t, x_4t, x_5t, x_6t, x_3x_5t, x_4x_6t]$. We define the multigrading by $\deg(x_1^{a_1} \ldots x_6^{a_6} t^\alpha) = (a_3, a_4, \alpha)$. Let $A = \{(0,0,1),(1,0,1),(0,1,1)\}$. Then, the toric fiber product $K[\mathcal{Q}_{G_1}] \times_A K[\mathcal{Q}_{G_2}]$ is generated by the monomials $x_1^{a_1} \ldots x_4^{a_4} t^\alpha \cdot x_3^{b_3} \ldots x_6^{b_6} t^\beta$ such that $x_1^{a_1} \ldots x_4^{a_4} t^\alpha \in K[\mathcal{Q}_{G_1}]$ and $x_3^{b_3} \ldots x_6^{b_6} t^\beta \in K[\mathcal{Q}_{G_2}]$ have the same multidegree $\mathbf{a} \in A$. It is easy to see that $K[\mathcal{Q}_{G_1}] \times_A K[\mathcal{Q}_{G_2}] \cong K[\mathcal{Q}_{G_1 \cup G_2}]$, and $A$ corresponds to the lattice point in $\mathcal{Q}_{G_1 \cap G_2}$, which is a simplex.*

The application of toric fiber products to toric rings of stable set polytopes was studied in [11]. For $i = 1, 2$, let $G_i$ be a simple graph on the vertex set $V_i$ and the edge set $E_i$. If $V_1 \cap V_2$ is a clique of both $G_1$ and $G_2$, then we construct a new graph $G_1 \sharp G_2$ on the vertex set $V_1 \cup V_2$ and the edge set $E_1 \cup E_2$, which is called the *clique sum* of $G_1$ and $G_2$ along $V_1 \cap V_2$.

**Proposition 1.** *Let $G_1 \sharp G_2$ be the clique sum of simple graphs $G_1$ and $G_2$. Then, $I_{\mathcal{Q}_{G_1 \sharp G_2}}$ is a toric fiber product of $I_{\mathcal{Q}_{G_1}}$ and $I_{\mathcal{Q}_{G_2}}$. We can construct a set of binomial generators (or a Gröbner basis) of $I_{\mathcal{Q}_{G_1 \sharp G_2}}$ from that of $I_{\mathcal{Q}_{G_i}}$'s and some quadratic binomials. Moreover, $K[\mathcal{Q}_{G_1 \sharp G_2}]$ is normal if and only if both $K[\mathcal{Q}_{G_1}]$ and $K[\mathcal{Q}_{G_2}]$ are normal.*

**Proof.** Note that $I_{\mathcal{Q}_{G_1 \cap G_2}} = \{0\}$. Hence, this is a special case of ([11] Proposition 5.1) by ([11] Proposition 9.6). □

A simple graph $G$ is called *chordal* if any induced cycle of $G$ is of length 3. A graph $G$ is called a *ring graph* if each block of $G$ that is not a bridge or a vertex can be constructed from a cycle by successively adding cycles of length $\geq 3$ using the clique sum construction. Ring graphs are introduced in [13,14].

**Proposition 2.** *Suppose that a simple graph $G$ is either a chordal graph or a ring graph. Then, $I_{\mathcal{Q}_G}$ has a squarefree quadratic initial ideal.*

**Proof.** It is known by ([15] Proposition 5.5.1) that a graph $G$ is chordal if and only if $G$ is a clique sum of complete graphs. By the statement in Example 1 and Proposition 1, $I_{\mathcal{Q}_G}$ has a squarefree quadratic initial ideal if $G$ is chordal.

Suppose that $G$ is a ring graph. Then, $G$ is a clique sum of trees and cycles since any graph is a clique sum (along a vertex) of trees and its blocks. Since trees and cycles are almost bipartite, by ([11] Theorem 8.1), the toric ideal $I_{\mathcal{Q}_H}$ has a squarefree quadratic initial ideal if $H$ is either a tree or a cycle. Thus, by Proposition 1, $I_{\mathcal{Q}_G}$ has a squarefree quadratic initial ideal if $G$ is a ring graph. □

A graph $G$ is called *perfect* if the chromatic number of every induced subgraph of $G$ is equal to the size of the largest clique of that subgraph; see [15]. We recall the following result on perfect graphs and their stable set polytopes; see [3,4,7].

**Proposition 3.** *Let $G$ be a simple graph. Then, $\mathcal{Q}_G$ is compressed if and only if $G$ is perfect. In particular, if $G$ is perfect, then $\mathcal{Q}_G$ is normal.*

For a graph $G$ on the vertex set $[n]$, let $\overline{G}$ denote the complement of a graph $G$. An induced cycle of $G$ of length $> 3$ is called a *hole* of $G$ and an induced cycle of $\overline{G}$ of length $> 3$ is called an *antihole* of $G$. Below we combine two important characterizations of perfect graphs, where the first part is the strong perfect graph theorem and the second part considers just the perfect graphs with stability number 2.

**Proposition 4.** *Let $G$ be a simple graph. Then $G$ is a perfect graph if and only if $G$ has no odd holes and no odd antiholes. In particular, $G$ is a perfect graph with $\alpha(G) = 2$ if and only if $\overline{G}$ is bipartite and not empty.*

For a graph $G$, let $\overline{G}^*$ be the nonsimple graph on the vertex set $[n+1]$ whose edge (and loop) set is $E(\overline{G}) \cup \{\{i, n+1\} : i \in [n+1]\}$. The following lemma plays an important role when we study the stable set polytope $\mathcal{Q}_G$ of $G$.

**Lemma 1.** *Let $G$ be a simple graph with $\alpha(G) = 2$. Then we have $K[\mathcal{Q}_G] \simeq K[\mathcal{P}_{\overline{G}^*}]$. Moreover, if $\overline{G}$ is bipartite, then there exists a bipartite graph $H$ such that $K[\mathcal{Q}_G] \simeq K[\mathcal{P}_H]$.*

**Proof.** Let $\varphi : K[\mathcal{Q}_G] \to K[x_1, \ldots, x_{n+1}]$ be the injective ring homomorphism defined by

$$\varphi(x_1^{a_1} \cdots x_n^{a_n} t) = x_1^{a_1} \cdots x_n^{a_n} x_{n+1}^{2-(a_1+\cdots+a_n)}.$$

Then $\varphi(t) = x_{n+1}^2$, $\varphi(x_i t) = x_i x_{n+1}$ for $i = 1, 2, \ldots, n$, and $\varphi(x_k x_\ell t) = x_k x_\ell$ for each stable set $\{k, \ell\}$ of $G$. Note that $\{k, \ell\}$ is a stable set of $G$ if and only if $\{k, \ell\}$ is an edge of $\overline{G}$. Hence, the image of $\varphi$ is $K[\mathcal{P}_{\overline{G}^*}]$.

Suppose that $\overline{G}$ is bipartite. Then $\overline{G}$ has no odd cycles. Hence, any odd cycle of $\overline{G}^*$ have the vertex $n+1$. (Note that $\{n+1, n+1\}$ is an odd cycle of length 1.) Thus, in particular, any two odd cycles of $\overline{G}^*$ has a common vertex. We now show that there exists a bipartite graph $H$ such that $K[\mathcal{P}_{\overline{G}^*}] \simeq K[\mathcal{P}_H]$ (by a similar argument in ([16] Proof of Proposition 5.5)). Let $[n] = V_1 \cup V_2$ be a partition of the vertex set of the bipartite graph $\overline{G}$. Let $H$ be a bipartite graph on the vertex set $[n+2]$ and the edge set

$$E(H) = E(\overline{G}) \cup \{\{i, n+1\} : i \in V_1\} \cup \{\{j, n+2\} : j \in V_2\} \cup \{\{n+1, n+2\}\}.$$

Let $\psi : K[\mathcal{P}_H] \to K[x_1, \ldots, x_{n+1}]$ be the ring homomorphism defined by

$$\psi(x_1^{a_1} \cdots x_{n+2}^{a_{n+2}}) = x_1^{a_1} \cdots x_n^{a_n} x_{n+1}^{a_{n+1}+a_{n+2}}.$$

Since $\overline{G}^*$ is obtained from $H$ by identifying the vertices $n+1$ and $n+2$, it follows that the image of $\psi$ is $K[\mathcal{P}_{\overline{G}^*}]$. Hence, it is enough to show that $\psi$ is injective. Suppose that $u = x_1^{a_1} \cdots x_{n+2}^{a_{n+2}}$, $v = x_1^{b_1} \cdots x_{n+2}^{b_{n+2}} \in K[\mathcal{P}_H]$ satisfies $\psi(u) = \psi(v)$. Then $a_{n+1} + a_{n+2} = b_{n+1} + b_{n+2}$ and $a_i = b_i$ for $i = 1, 2, \ldots, n$. Since $[n] = V_1 \cup V_2$ is a partition for $\overline{G}$, we have

$$a_{n+2} + \sum_{k \in V_1} a_k = a_{n+1} + \sum_{\ell \in V_2} a_\ell, \quad b_{n+2} + \sum_{k \in V_1} b_k = b_{n+1} + \sum_{\ell \in V_2} b_\ell.$$

Thus, $a_{n+1} = b_{n+1}$ and $a_{n+2} = b_{n+2}$, as desired. □

The first application of Lemma 1 is as follows:

**Proposition 5.** *Let $G$ be a simple graph. Then the following conditions are equivalent:*

(i) $\mathcal{Q}_G$ *is unimodular;*
(ii) $\overline{G}$ *is bipartite.*

*Moreover, if $\alpha(G) = 2$, then the conditions*

(iii) $\mathcal{Q}_G$ *is compressed;*
(iv) $G$ *is perfect*

*are also equivalent to conditions (i) and (ii).*

**Proof.** We may assume that $G$ is not complete (i.e., $\overline{G}$ is not empty and $\alpha(G) \neq 1$). Let $A$ be the matrix whose columns are vertices of $\mathcal{Q}_G$, and let

$$\widetilde{A} = \begin{pmatrix} & A & \\ 1 & \cdots & 1 \end{pmatrix} \quad \text{and} \quad B = \begin{pmatrix} 0 & e_1 & \cdots & e_n \\ 1 & 1 & \cdots & 1 \end{pmatrix}.$$

Then $B$ is a submatrix of $\widetilde{A}$. Since $|\det(B)| = 1$, the rank of $\widetilde{A}$ is $n+1$. It is known by ([1] p. 70) that $\mathcal{Q}_G$ is unimodular if and only if the absolute value of any nonzero $(n+1)$-minor of the matrix $\widetilde{A}$ is 1.

Suppose that $\overline{G}$ is not bipartite. Then $\overline{G}$ has an odd cycle $C = (i_1, \ldots, i_{2\ell+1})$. Then the absolute value of the $(n+1)$-minor of $\widetilde{A}$ that corresponds to

$$\{e_{i_k} + e_{i_{k+1}} : 1 \leq k \leq 2\ell\} \cup \{e_{i_1} + e_{i_{2\ell+1}}, 0\} \cup \{e_j : j \notin \{i_1, \ldots, i_{2\ell+1}\}\}$$

equals 2. Hence, $\mathcal{Q}_G$ is not unimodular. Thus, we have (i) ⇒ (ii).

Suppose that $\overline{G}$ is bipartite. By Lemma 1, there exists a bipartite graph $H$ such that $K[\mathcal{Q}_G] \simeq K[\mathcal{P}_H]$. It is well known that the edge polytope of a bipartite graph is unimodular; see ([2] Theorem 5.24). Thus, $\mathcal{P}_H$ is unimodular, and hence we have (ii) ⇒ (i).

Suppose that $\alpha(G) = 2$. By Proposition 3, conditions (iii) and (iv) are equivalent. In addition, by Proposition 4, conditions (ii) and (iv) are equivalent. □

We close this section with the following fundamental fact on stable set and edge polytopes.

**Proposition 6.** *Let $G'$ be an induced subgraph of a graph $G$. Then*

(i) *the edge polytope $\mathcal{P}_{G'}$ is a face of $\mathcal{P}_G$;*
(ii) *if $G$ is a simple graph, then $\mathcal{Q}_{G'}$ is a face of $\mathcal{Q}_G$.*

It can be seen from Proposition 6 that several properties of $K[\mathcal{Q}_G]$ (resp. $K[\mathcal{P}_G]$) are inherited to $K[\mathcal{Q}_{G'}]$ (resp. $K[\mathcal{P}_{G'}]$), for example, normality of the toric ring, the existence of a squarefree initial ideal, existence of a quadratic Gröbner basis, and the existence of the set of quadratic binomial generators of the toric ideal; see [17].

## 3. Normality of Stable Set Polytopes

In this section, we study the normality of stable set polytopes. Normality of edge polytopes is studied in [18,19], and we make use of the normality conditions of edge polytopes while working with stable set polytopes, due to the relation discovered in Lemma 1. If $C_1$ and $C_2$ are cycles in a graph $G$, then $\{i,j\} \in E(G)$ is called a *bridge* of $C_1$ and $C_2$ if $i$ is a vertex of $C_1 \setminus C_2$ and $j$ is a vertex of $C_2 \setminus C_1$. We say that a graph $G$ satisfies the *odd cycle condition* if any induced odd cycles $C_1$ and $C_2$ in $G$ have either a common vertex or a bridge. For the sake of simplicity, assume that a graph $H$ has at most one loop. Then, it is known by [18,19] that $\mathcal{P}_H$ is normal if and only if each connected component of $H$ satisfies the odd cycle condition. By ([18] Corollary 2.3) and Lemma 1, we have the following. (Note that $\overline{G}$ below is not necessarily connected.)

**Theorem 1.** *Let $G$ be a simple graph with $\alpha(G) = 2$. Then the following conditions are equivalent.*

(i) $\mathcal{Q}_G$ is normal;
(ii) $\mathcal{Q}_G$ has a unimodular covering;
(iii) $\overline{G}$ satisfies the odd cycle condition, i.e., if two odd holes $C_1$ and $C_2$ in $\overline{G}$ have no common vertices, then there exists a bridge of $C_1$ and $C_2$ in $\overline{G}$.

*In particular, if $\mathcal{Q}_G$ is normal, then $\mathcal{P}_{\overline{G}}$ is normal.*

**Proof.** By Lemma 1, we have $K[\mathcal{Q}_G] \simeq K[\mathcal{P}_{\overline{G}^*}]$. Hence, by ([18] Corollary 2.3), conditions (i) and (ii) are equivalent, and they hold if and only if $\overline{G}^*$ satisfies the odd cycle condition. (Note that $\overline{G}^*$ is connected.) Since the vertex $n+1$ is incident to any vertex of $\overline{G}^*$, it is easy to see that $\overline{G}^*$ satisfies the odd cycle condition if and only if $\overline{G}$ satisfies the odd cycle condition. □

It is shown in [20] that there exists a graph $G$ such that $\mathcal{P}_G$ is normal and that $I_{\mathcal{P}_G}$ has no squarefree initial ideals. Examples on infinite families of such edge polytopes are given in [21]. We can construct the stable set polytopes with the same properties. Let $G_1(p_1,\ldots,p_5)$ be the graph defined in ([21] Theorem 3.10).

**Theorem 2.** *Let $G$ be a graph such that $\overline{G} = G_1(p_1,\ldots,p_5)$ with $p_i \geq 2$ for $i = 1,\ldots,5$. Then $\mathcal{Q}_G$ is normal, and $I_{\mathcal{Q}_G}$ has no squarefree initial ideals.*

**Proof.** Since $\overline{G}$ has no triangles, we have $\alpha(G) = 2$. Since $G_1(p_1,\ldots,p_5)$ satisfies the odd cycle condition, $\mathcal{Q}_G$ is normal by Theorem 1. On the other hand, $I_{\overline{G}}$ has no squarefree initial ideals. Since $\overline{G}$ is an induced subgraph of $\overline{G}^*$, $I_{\mathcal{Q}_G}$ has no squarefree initial ideals by Lemma 1 and Proposition 6. □

It seems to be a challenging problem to characterize the normal stable set polytopes with large stability numbers. We give several necessary conditions. The following is a consequence of Proposition 6 and Theorem 1.

**Proposition 7.** *Let $G$ be a simple graph. Suppose that $\mathcal{Q}_G$ is normal. Then any two odd holes of $\overline{G}$ without a common vertex have a bridge in $\overline{G}$.*

**Proof.** Suppose that two odd holes $C_1, C_2$ of $\overline{G}$ without common vertices have no bridges in $\overline{G}$. Let $H$ be an induced subgraph of $G$ whose vertex set is that of $C_1 \cup C_2$. Then $\alpha(H) = 2$, and hence $\mathcal{Q}_H$ is not normal by Theorem 1. Thus, $\mathcal{Q}_G$ is not normal by Proposition 6. □

Similar conditions are required for antiholes of $\overline{G}$.

**Proposition 8.** *Let $G$ be a simple graph. Suppose that $\mathcal{Q}_G$ is normal. Then $G$ satisfies all of the following conditions:*

(i) *Any two odd antiholes of $\overline{G}$ having no common vertices have a bridge in $\overline{G}$.*
(ii) *Any two odd antiholes of $\overline{G}$ of length $\geq 7$ having exactly one common vertex have a bridge in $\overline{G}$.*
(iii) *Any odd hole and odd antihole of $\overline{G}$ having no common vertices have a bridge in $\overline{G}$.*

**Proof.** Let $G$ be a graph on the vertex set $[n]$. Let $\mathcal{A} = \{(\rho(W),1) : W \in S(G)\}$. It is known by ([1] Proposition 13.5) that $\mathcal{Q}_G$ is normal if and only if we have $\mathbb{Z}_{\geq 0}\mathcal{A} = \mathbb{Q}_{\geq 0}\mathcal{A} \cap \mathbb{Z}^{n+1}$.

(i) Let $C_1 = (i_1, \ldots, i_{2k+1})$ and $C_2 = (j_1, \ldots, j_{2\ell+1})$ be odd antiholes in $\overline{G}$ having no common vertices and no bridges in $\overline{G}$. By Proposition 6, we may assume that $\overline{G} = C_1 \cup C_2$. Then,

$$\sum_{W \in S(\overline{C_1}) \text{ and } |W|=k} (\rho(W),1) = (2k+1)\mathbf{e}_{n+1} + k\sum_{p=1}^{2k+1} \mathbf{e}_{i_p},$$

$$\sum_{W \in S(\overline{C_2}) \text{ and } |W|=\ell} (\rho(W),1) = (2\ell+1)\mathbf{e}_{n+1} + \ell\sum_{q=1}^{2\ell+1} \mathbf{e}_{j_q}.$$

Since $k, \ell \geq 2$, we have $k\ell - k - \ell \geq 0$. Hence,

$$\begin{aligned}
\alpha &:= 5\mathbf{e}_{n+1} + \sum_{p=1}^{2k+1} \mathbf{e}_{i_p} + \sum_{q=1}^{2\ell+1} \mathbf{e}_{j_q} \\
&= \frac{1}{k}\left((2k+1)\mathbf{e}_{n+1} + k\sum_{p=1}^{2k+1} \mathbf{e}_{i_p}\right) + \frac{1}{\ell}\left((2\ell+1)\mathbf{e}_{n+1} + \ell\sum_{q=1}^{2\ell+1} \mathbf{e}_{j_q}\right) + \frac{k\ell - k - \ell}{k\ell}\mathbf{e}_{n+1}
\end{aligned}$$

belongs to $\mathbb{Q}_{\geq 0}\mathcal{A} \cap \mathbb{Z}^{n+1}$. Suppose that $\alpha$ belongs to $\mathbb{Z}_{\geq 0}\mathcal{A}$. Since the $(n+1)$-th coordinate of $\alpha$ is 5, there exist $W_1, \ldots, W_5$ such that

$$\alpha = (\rho(W_1),1) + \cdots + (\rho(W_5),1),$$

where each $W_i$ belongs to either $S(\overline{C_1})$ or $S(\overline{C_2})$. It then follows that $\sum_{W_i \in S(\overline{C_1})} |W_i| = 2k+1$ and $\sum_{W_i \in S(\overline{C_2})} |W_i| = 2\ell+1$. Since $\max\{|W| : W \in S(\overline{C_1})\} = k$, $\max\{|W| : W \in S(\overline{C_2})\} = \ell$, we have

$$|\{W_1, \ldots, W_5\} \cap S(\overline{C_1})| \geq 3,$$
$$|\{W_1, \ldots, W_5\} \cap S(\overline{C_2})| \geq 3.$$

This is a contradiction. Thus, $\alpha$ is not in $\mathbb{Z}_{\geq 0}\mathcal{A}$.

(ii) Let $C_1 = (i_1, \ldots, i_{2k+1})$ and $C_2 = (j_1, \ldots, j_{2\ell+1})$ be odd antiholes in $\overline{G}$ of length $\geq 7$ having exactly one common vertex $i_1 = j_1$ and no bridges in $\overline{G}$. By Proposition 6, we may assume that $\overline{G} = C_1 \cup C_2$. Let

$$\begin{aligned}
S_1 &= \{W \in S(\overline{C_1}) : |W| = k \text{ and either } i_1 \in W \text{ or } \{i_2, i_{2k+1}\} \subset W\}, \\
S_2 &= \{W \in S(\overline{C_2}) : |W| = \ell \text{ and either } j_1 \in W \text{ or } \{j_2, j_{2\ell+1}\} \subset W\}.
\end{aligned}$$

Then,

$$\sum_{W \in S_1} (\rho(W), 1) = (2k-1)\mathbf{e}_{n+1} + k\mathbf{e}_{i_1} + (k-1) \sum_{p=2}^{2k+1} \mathbf{e}_{i_p},$$

$$\sum_{W \in S_2} (\rho(W), 1) = (2\ell-1)\mathbf{e}_{n+1} + \ell\mathbf{e}_{i_1} + (\ell-1) \sum_{q=2}^{2\ell+1} \mathbf{e}_{j_q}.$$

Since $k, \ell \geq 3$, we have $0 < 1/(k-1) + 1/(\ell-1) \leq 1$. Hence,

$$\alpha := 5\mathbf{e}_{n+1} + 3\mathbf{e}_{i_1} + \sum_{p=2}^{2k+1} \mathbf{e}_{i_p} + \sum_{q=2}^{2\ell+1} \mathbf{e}_{j_q}$$

$$= \frac{1}{k-1} \left( (2k-1)\mathbf{e}_{n+1} + k\mathbf{e}_{i_1} + (k-1) \sum_{p=2}^{2k+1} \mathbf{e}_{i_p} \right)$$

$$+ \frac{1}{\ell-1} \left( (2\ell-1)\mathbf{e}_{n+1} + \ell\mathbf{e}_{i_1} + (\ell-1) \sum_{q=2}^{2\ell+1} \mathbf{e}_{j_q} \right) + \left( 1 - \frac{1}{k-1} - \frac{1}{\ell-1} \right) (\mathbf{e}_{i_1} + \mathbf{e}_{n+1})$$

belongs to $\mathbb{Q}_{\geq 0} \mathcal{A} \cap \mathbb{Z}^{n+1}$. Since the $(n+1)$-th coordinate of $\alpha$ is 5, there exist $W_1, \ldots, W_5 \in S(\overline{G})$ such that $\alpha = (\rho(W_1), 1) + \cdots + (\rho(W_5), 1)$. Then, each $W_i$ belongs to either $S(\overline{C_1})$ or $S(\overline{C_2})$. Since $\max\{|W| : W \in S(\overline{C_1})\} = k$, $\max\{|W| : W \in S(\overline{C_2})\} = \ell$, we have

$$|\{W_1, \ldots, W_5\} \cap S(\overline{C_1})| \geq 2,$$
$$|\{W_1, \ldots, W_5\} \cap S(\overline{C_2})| \geq 2.$$

Thus, $(|\{W_1, \ldots, W_5\} \cap S(\overline{C_1})|, |\{W_1, \ldots, W_5\} \cap S(\overline{C_2})|)$ is either $(2, 3)$ or $(3, 2)$. Changing indices if necessary, we may assume that $W_1, W_2 \in S(\overline{C_1})$ and $W_3, W_4, W_5 \in S(\overline{C_2})$. It then follows that $\rho(W_1) + \rho(W_2) = \sum_{p=2}^{2k+1} \mathbf{e}_{i_p}$, and hence $\rho(W_3) + \rho(W_4) + \rho(W_5) = 3\mathbf{e}_{i_1} + \sum_{q=2}^{2\ell+1} \mathbf{e}_{j_q}$. This implies that $i_1 \in W_3 \cap W_4 \cap W_5$. Thus, $i_2, i_{2\ell+1} \notin W_3, W_4, W_5$, a contradiction. Therefore, we have $\alpha \notin \mathbb{Z}_{\geq 0} \mathcal{A}$.

(iii) Let $C_1 = (i_1, \ldots, i_{2k+1})$ be an odd hole and $C_2 = (j_1, \ldots, j_{2\ell+1})$ an odd antihole in $\overline{G}$ having no common vertices. By Proposition 6, we may assume that $\overline{G} = C_1 \cup C_2$. Then,

$$\sum_{W \in S(\overline{C_1}) \text{ and } |W|=2} (\rho(W), 1) = (2k+1)\mathbf{e}_{n+1} + 2 \sum_{p=1}^{2k+1} \mathbf{e}_{i_p},$$

$$\sum_{W \in S(\overline{C_2}) \text{ and } |W|=\ell} (\rho(W), 1) = (2\ell+1)\mathbf{e}_{n+1} + \ell \sum_{q=1}^{2\ell+1} \mathbf{e}_{j_q}.$$

Hence,

$$(k+3)\mathbf{e}_{n+1} + \sum_{p=1}^{2k+1} \mathbf{e}_{i_p} + \sum_{q=1}^{2\ell+1} \mathbf{e}_{j_q}$$

$$= \frac{1}{2} \left( (2k+1)\mathbf{e}_{n+1} + 2 \sum_{p=1}^{2k+1} \mathbf{e}_{i_p} \right) + \frac{1}{\ell} \left( (2\ell+1)\mathbf{e}_{n+1} + \ell \sum_{q=1}^{2\ell+1} \mathbf{e}_{j_q} \right) + \frac{\ell - 2}{2\ell} \mathbf{e}_{n+1}$$

belongs to $\mathbb{Q}_{\geq 0} \mathcal{A} \cap \mathbb{Z}^{n+1}$. However, this vector is not in $\mathbb{Z}_{\geq 0} \mathcal{A}$ since $\max\{|W| : W \in S(\overline{C_1})\} = 2$, $\max\{|W| : W \in S(\overline{C_2})\} = \ell$, and $\lceil (2k+1)/2 \rceil + \lceil (2\ell+1)/\ell \rceil = k + 4 > k + 3$. □

Unfortunately, the above conditions are not sufficient to be normal in general. For example, if the length of the two odd antiholes of $\overline{G}$ without common vertices are long, then a lot of bridges in $\overline{G}$ seem to be needed.

## 4. Generators and Gröbner Bases of $I_{\mathcal{Q}_G}$

For a toric ideal $I$, let $d(I)$ be the maximum degree of binomials in a minimal set of binomial generators of $I$. If $I = \{0\}$, then we set $d(I) = 0$. In this section, we study $d(I_{\mathcal{Q}_G})$ by using results on the toric ideals of edge polytopes.

Let $G$ be a graph on the vertex set $[n]$ allowing loops and having no multiple edges. Let $E(G) = \{e_1, \ldots, e_m\}$ be a set of all edges and loops of $G$. The toric ideal $I_{\mathcal{P}_G}$ is the kernel of a homomorphism $\pi : K[y_1, \ldots, y_m] \to K[x_1, \ldots, x_n]$ defined by $\pi(y_i) = x_k x_\ell$ where $e_i = \{k, \ell\}$. A finite sequence of the form

$$\Gamma = (\{v_1, v_2\}, \{v_2, v_3\}, \ldots, \{v_q, v_{q+1}\}) \tag{1}$$

with each $\{v_k, v_{k+1}\} \in E(G)$ is called a *walk* of length $q$ of $G$ connecting $v_1 \in [n]$ and $v_{q+1} \in [n]$. A walk $\Gamma$ of the form (1) is called *even* (resp. *odd*) if $q$ is even (resp. odd). A walk $\Gamma$ of the form (1) is called *closed* if $v_{q+1} = v_1$. Given an even closed walk $\Gamma = (e_{i_1}, e_{i_2}, \ldots, e_{i_{2q}})$ of $G$, we write $f_\Gamma$ for the binomial

$$f_\Gamma = \prod_{k=1}^{q} y_{i_{2k-1}} - \prod_{k=1}^{q} y_{i_{2k}} \in I_{\mathcal{P}_G}.$$

We regard a loop as an odd cycle of length 1. We recall the following result from [1,5,22].

**Proposition 9.** *Let $G$ be a graph having at most one loop. Then $I_{\mathcal{P}_G}$ is generated by all the binomials $f_\Gamma$, where $\Gamma$ is an even closed walk of $G$. In particular, $I_{\mathcal{P}_G} = (0)$ if and only if each connected component of $G$ has at most one cycle and the cycle is odd.*

The following theorem implies that the set of binomial generators of $I_{\mathcal{Q}_G}$ can also be characterized by the graph-theoretical terminology if $\alpha(G) = 2$.

**Theorem 3.** *Let $G$ be a simple graph with $\alpha(G) = 2$. Then, $I_{\mathcal{Q}_G} = I_{\mathcal{P}_{\overline{G}}} + J$ where $J$ is an ideal generated by quadratic binomials $f_\Gamma$ where $\Gamma$ is an even closed walk of $\overline{G}^\star$ that satisfies one of the following:*

(i)  $\Gamma = (\{i,j\},\{j,k\},\{k,n+1\},\{n+1,i\})$ *is a cycle where* $\{i,j\}, \{j,k\} \in E(\overline{G})$;
(ii) $\Gamma = (\{i,j\},\{j,n+1\},\{n+1,n+1\},\{n+1,i\})$ *where* $\{i,j\} \in E(\overline{G})$.

*In particular, $d(I_{\mathcal{Q}_G}) = \max\{d(I_{\mathcal{P}_{\overline{G}}}), 2\}$.*

**Proof.** Since $\alpha(G) = 2$, we have $I_{\mathcal{Q}_G} = I_{\mathcal{P}_{\overline{G}^\star}}$ by Lemma 1. Since $\overline{G}$ is a subgraph of $\overline{G}^\star$, it follows that $I_{\mathcal{P}_{\overline{G}^\star}} \supset I_{\mathcal{P}_{\overline{G}}} + J$. Thus, it is enough to show that $I_{\mathcal{P}_{\overline{G}^\star}} \subset I_{\mathcal{P}_{\overline{G}}} + J$.

Let $\Gamma$ be an even closed walk of $\overline{G}^\star$. It is enough to show that $f_\Gamma \in I_{\mathcal{P}_{\overline{G}^\star}}$ belongs to $I_{\mathcal{P}_{\overline{G}}} + J$. Suppose that $f_\Gamma$ does not belong to $I_{\mathcal{P}_{\overline{G}}} + J$. Then, the vertex $n+1$ belongs to $\Gamma$. We may assume that the degree of $f_\Gamma$ is minimum among binomials in $I_{\mathcal{P}_{\overline{G}^\star}}$ that do not belong to $I_{\mathcal{P}_{\overline{G}}} + J$. Then, $f_\Gamma$ is irreducible. Let

$$\Gamma = (\{n+1, p\}, \{p, q\}, \{q, r\}, \Gamma'),$$

where $\Gamma' = (e_{i_1}, \ldots, e_{i_{2k+1}})$ is an odd subwalk of $\Gamma$ from the vertex $r$ to the vertex $n+1$. Since $f_\Gamma$ is irreducible, it follows that $p, q, r$ are distinct vertices and that $q \neq n+1$. Then,

$$f_\Gamma = \left(\prod_{\alpha=1}^{k} y_{i_{2\alpha}}\right) f_{\Gamma_1} + y_j f_{\Gamma_2},$$

where $e_j = \{p, q\}$, $\Gamma_1 = (\{n+1, p\}, \{p, q\}, \{q, r\}, \{r, n+1\})$, and $\Gamma_2 = (\{n+1, r\}, \Gamma')$. Since $\deg f_{\Gamma_2} < \deg f_\Gamma$, the binomial $f_{\Gamma_2}$ belongs to $I_{\mathcal{P}_{\overline{G}}} + J$ by the assumption on $\deg f_\Gamma$. Moreover, $\Gamma_1$ satisfies one of the conditions (i) or (ii), and hence $f_{\Gamma_1} \in J$. Thus, we have $f_\Gamma \in \langle f_{\Gamma_1}, f_{\Gamma_2} \rangle \subset I_{\mathcal{P}_{\overline{G}}} + J$, a contradiction. □

**Remark 1.** *A graph-theoretical characterization of a simple graph G such that $d(I_{\mathcal{P}_G}) \leq 2$ is given in [5]. Then, by making use of Theorem 3, one can provide a similar characterization of a simple graph G where $\alpha(G) = 2$ and $d(I_{\mathcal{Q}_G}) \leq 2$.*

It is known by ([11] Theorem 7.4) that if the complement of a graph G is an even cycle of length $2k$, then we have $d(I_{\mathcal{Q}_G}) = k$. By Theorem 3, we can generalize this result for a graph whose complement is an arbitrary bipartite graph.

**Corollary 1.** *Let G be a simple graph such that $\overline{G}$ is bipartite. Then we have*

$$d(I_{\mathcal{Q}_G}) = \begin{cases} 0 & \text{if } \overline{G} \text{ is empty (i.e., G is complete)}, \\ k & \text{if } \overline{G} \text{ has a cycle}, \\ 2 & \text{otherwise}, \end{cases}$$

*where $2k$ is the maximum length of induced cycles of $\overline{G}$. Moreover, the following conditions are equivalent:*

(i) $d(I_{\mathcal{Q}_G}) \leq 2$, i.e., $I_{\mathcal{Q}_G}$ is generated by quadratic binomials;
(ii) $K[\mathcal{Q}_G]$ is Koszul;
(iii) $I_{\mathcal{Q}_G}$ has a quadratic Gröbner basis;
(iv) the length of any induced cycle of $\overline{G}$ is 4.

**Proof.** Let G be a simple graph such that $\overline{G}$ is bipartite. Then $\alpha(G) = 2$. Since $\overline{G}$ is bipartite, it is known (see [23] Lemma 2.4) that $I_{\mathcal{P}_{\overline{G}}}$ is generated by $f_\Gamma$ where $\Gamma$ is an induced even cycle of $\overline{G}$. Note that $\deg f_\Gamma = k$ if the length of $\Gamma$ is $2k$. Hence, by Theorem 3, we obtain the desired formula for $d(I_{\mathcal{Q}_G})$.

It follows from the formula of $d(I_{\mathcal{Q}_G})$ that (i) and (iv) are equivalent. Moreover, (iii) $\Rightarrow$ (ii) $\Rightarrow$ (i) holds in general. By Lemma 1, there exists a bipartite graph H such that $K[\mathcal{Q}_G] \simeq K[\mathcal{P}_H]$. By ([24] Theorem), $I_{\mathcal{P}_H}$ has a quadratic Gröbner basis if and only if $d(I_{\mathcal{P}_H}) \leq 2$. Thus, we have (i) $\Rightarrow$ (iii). □

If $\overline{G}$ is not bipartite, then condition (i) and (iii) in Corollary 1 are not equivalent. In order to construct an infinite family of counterexamples, the following proposition is important. (Proof is essentially given in ([5] Proof of Proposition 1.6)).

**Proposition 10.** *Let $\mathcal{P}$ be a $(0,1)$-polytope. If $I_\mathcal{P}$ has a quadratic Gröbner basis, then the initial ideal is generated by squarefree monomials, and hence $\mathcal{P}$ is normal.*

**Theorem 4.** *Let G be a simple graph such that $\overline{G} = C_1 \cup C_2$, where $C_1$ and $C_2$ are odd holes without common vertices. Then $\alpha(G) = 2$, and*

(a) $I_{\mathcal{Q}_G}$ is generated by quadratic binomials;
(b) $I_{\mathcal{Q}_G}$ has no quadratic Gröbner bases;
(c) $\mathcal{Q}_G$ is not normal.

**Proof.** Since $\overline{G}$ has no triangles, we have $\alpha(G) = 2$. Since each connected component of $\overline{G}$ is an odd cycle, $I_{\mathcal{P}_{\overline{G}}} = \{0\}$ by Proposition 9. It follows from Theorem 3 that $I_{\mathcal{Q}_G}$ is generated by quadratic binomials. By Theorem 1, $\mathcal{Q}_G$ is not normal since $C_1$ and $C_2$ have no bridges. Thus, by Proposition 10, $I_{\mathcal{Q}_G}$ has no quadratic Gröbner bases. □

The graphs in Theorem 4 are not strongly Koszul by ([9] Theorem 1.3). However, we do not know whether they are Koszul or not in general.

**Remark 2.** *It is known by ([5] Theorem 1.2) that if G is a simple connected graph and $I_{\mathcal{P}_G}$ is generated by quadratic binomials, then G satisfies the odd cycle condition, and hence $\mathcal{P}_G$ is normal.*

It seems to be a challenging problem to characterize the graphs $G$ such that $\alpha(G) > 2$ and $d(I_{Q_G}) \leq 2$. The following is a consequence of Proposition 6, Theorem 3 and ([5] Theorem 1.2).

**Proposition 11.** *Let $G$ be a simple graph. If $I_{Q_G}$ is generated by quadratic binomials, then $\overline{G}$ satisfies the following conditions:*

(i) *Any even cycle of $\overline{G}$ of length $\geq 6$ has a chord;*
(ii) *Any two odd holes of $\overline{G}$ having exactly one common vertex have a bridge;*
(iii) *Any two odd holes of $\overline{G}$ having no common vertex have at least two bridges.*

**Proof.** Suppose that $\overline{G}$ does not satisfy one of the conditions above. If $\overline{G}$ does not satisfy condition (i), then let $H$ be an induced subgraph of $G$ whose vertex set is that of the even cycle. If $\overline{G}$ does not satisfy either condition (ii) or (iii), then let $H$ be an induced subgraph of $G$ whose vertex set is that of two odd holes. Then $\alpha(H) = 2$, and hence $I_{Q_H}$ is not generated by quadratic binomials by Theorem 3 and ([5] Theorem 1.2). Thus, it follows from Proposition 6 that $I_{Q_G}$ is not generated by quadratic binomials. □

**Author Contributions:** All authors made equal and significant contributions to writing this article, and approved the final manuscript.

**Funding:** The authors were partially supported by JSPS KAKENHI 17K14165 and 18H01134.

**Acknowledgments:** The authors are grateful to the anonymous referees for their careful reading and helpful comments.

**Conflicts of Interest:** The authors declare no conflict of interest.

### References

1. Sturmfels, B. *Gröbner Bases and Convex Polytopes*; American Mathematical Society: Providence, RI, USA, 1996.
2. Herzog, J.; Hibi, T.; Ohsugi, H. *Binomial Ideals*; Graduate Texts in Math 279; Springer International Publishing: Cham, Switzerland, 2018.
3. Ohsugi, H.; Hibi, T. Convex polytopes all of whose reverse lexicographic initial ideals are squarefree. *Proc. Am. Math. Soc.* **2001**, *129*, 2541–2546. [CrossRef]
4. Sullivant, S. Compressed polytopes and statistical disclosure limitation. *Tohoku Math. J.* **2006**, *58*, 433–445. [CrossRef]
5. Ohsugi, H.; Hibi, T. Toric ideals generated by quadratic binomials. *J. Algebra* **1999**, *218*, 509–527. [CrossRef]
6. Sturmfels, B. Four counterexamples in combinatorial algebraic geometry. *J. Algebra* **2000**, *230*, 282–294. [CrossRef]
7. Gouveia, J.; Parrilo, P.A.; Thomas, R.R. Theta bodies for polynomial ideals. *SIAM J. Optim.* **2010**, *20*, 2097–2118. [CrossRef]
8. Ohsugi, H.; Hibi, T. Special simplices and Gorenstein toric rings. *J. Combin. Theory Ser. A* **2006**, *113*, 718–725. [CrossRef]
9. Matsuda, K. Strong Koszulness of toric rings associated with stable set polytopes of trivially perfect graphs. *J. Algebra Appl.* **2014**, *13*, 1350138. [CrossRef]
10. Hibi, T.; Li, N. Chain polytopes and algebras with straightening laws. *Acta Math. Vietnam.* **2015**, *40*, 447–452. [CrossRef]
11. Engström, A.; Norén, P. Ideals of graph homomorphisms. *Ann. Comb.* **2013**, *17*, 71–103. [CrossRef]
12. Sullivant, S. Toric fiber products. *J. Algebra* **2007**, *316*, 560–577. [CrossRef]
13. Gitler, I.; Reyes, E.; Villarreal, R.H. Ring graphs and toric ideals. *Electron. Notes Discrete Math.* **2007**, *28*, 393–400. [CrossRef]
14. Gitler, I.; Reyes, E.; Villarreal, R.H. Ring graphs and complete intersection toric ideals. *Discrete Math.* **2010**, *310*, 430–441. [CrossRef]
15. Diestel, R. *Graph Theory*, 4th ed.; Graduate Texts in Mathematics 173; Springer International Publishing: Berlin, Germany, 2010.
16. Ohsugi, H.; Hibi, T. Centrally symmetric configurations of integer matrices. *Nagoya Math. J.* **2014**, *216*, 153–170. [CrossRef]

17. Ohsugi, H.; Herzog, J.; Hibi, T. Combinatorial pure subrings. *Osaka J. Math.* **2000**, *37*, 745–757.
18. Ohsugi, H.; Hibi, T. Normal polytopes arising from finite graphsn. *J. Algebra* **1998**, *207*, 409–426. [CrossRef]
19. Simis, A.; Vasconcelos, W.V.; Villarreal, R.H. The integral closure of subrings associated to graphs. *J. Algebra* **1998**, *199*, 281–289. [CrossRef]
20. Ohsugi, H.; Hibi, T. A normal $(0,1)$-polytope none of whose regular triangulations is unimodular. *Discrete Comput. Geom.* **1999**, *21*, 201–204. [CrossRef]
21. Ohsugi, H. Toric ideals and an infinite family of normal $(0,1)$-polytopes without unimodular regular triangulations. *Discrete Comput. Geom.* **2002**, *27*, 551–565. [CrossRef]
22. Villarreal, R. Rees algebras of edge ideals. *Commun. Algebra* **1995**, *23*, 3513–3524. [CrossRef]
23. Simis, A. On the Jacobian module associated to a graph. *Proc. Am. Math. Soc.* **1998**, *126*, 989–997. [CrossRef]
24. Ohsugi, H.; Hibi, T. Koszul bipartite graphs. *Adv. Appl. Math.* **1999**, *22*, 25–28. [CrossRef]

© 2019 by the authors. Licensee MDPI, Basel, Switzerland. This article is an open access article distributed under the terms and conditions of the Creative Commons Attribution (CC BY) license (http://creativecommons.org/licenses/by/4.0/).

Article

# Faces of 2-Dimensional Simplex of Order and Chain Polytopes

**Aki Mori**

Department of Pure and Applied Mathematics, Graduate School of Information Science and Technology, Osaka University, Suita, Osaka 565-0871, Japan; u035543c@alumni.osaka-u.ac.jp

Received: 11 August 2019; Accepted: 10 September 2019; Published: 14 September 2019

**Abstract:** Each of the descriptions of vertices, edges, and facets of the order and chain polytope of a finite partially ordered set are well known. In this paper, we give an explicit description of faces of 2-dimensional simplex in terms of vertices. Namely, it will be proved that an arbitrary triangle in 1-skeleton of the order or chain polytope forms the face of 2-dimensional simplex of each polytope. These results mean a generalization in the case of 2-faces of the characterization known in the case of edges.

**Keywords:** order polytope; chain polytope; partially ordered set

**MSC:** primary: 52B05; secondary: 06A07

## 1. Introduction

The combinatorial structure of the order polytope $\mathcal{O}(P)$ and the chain polytope $\mathcal{C}(P)$ of a finite poset (partially ordered set) $P$ is explicitly discussed in [1]. Moreover, in [2], the problem when the order polytope $\mathcal{O}(P)$ and the chain polytope $\mathcal{C}(P)$ are unimodularly equivalent is solved. It is also proved that the number of edges of the order polytope $\mathcal{O}(P)$ is equal to that of the chain polytope $\mathcal{C}(P)$ in [3]. In the present paper we give an explicit description of faces of 2-dimensional simplex of $\mathcal{O}(P)$ and $\mathcal{C}(P)$ in terms of vertices. In other words, we show that triangles in 1-skeleton of $\mathcal{O}(P)$ or $\mathcal{C}(P)$ are in one-to-one correspondence with faces of 2-dimensional simplex of each polytope. These results are a direct generalizations of [4] (Lemma 4, Lemma 5).

## 2. Definition and Known Results

Let $P = \{x_1, \ldots, x_d\}$ be a finite poset. To each subset $W \subset P$, we associate $\rho(W) = \sum_{i \in W} \mathbf{e}_i \in \mathbb{R}^d$, where $\mathbf{e}_1, \ldots, \mathbf{e}_d$ are the canonical unit coordinate vectors of $\mathbb{R}^d$. In particular $\rho(\varnothing)$ is the origin of $\mathbb{R}^d$. A *poset ideal* of $P$ is a subset $I$ of $P$ such that, for all $x_i$ and $x_j$ with $x_i \in I$ and $x_j \leqslant x_i$, one has $x_j \in I$. An *antichain* of $P$ is a subset $A$ of $P$ such that $x_i$ and $x_j$ belonging to $A$ with $i \neq j$ are incomparable. The empty set $\varnothing$ is a poset ideal as well as an antichain of $P$. We say that $x_j$ *covers* $x_i$ if $x_i < x_j$ and $x_i < x_k < x_j$ for no $x_k \in P$. A chain $x_{j_1} < x_{j_2} < \cdots < x_{j_\ell}$ of $P$ is called *saturated* if $x_{j_q}$ covers $x_{j_{q-1}}$ for $1 < q \leqslant \ell$. A *maximal chain* is a saturated chain such that $x_{j_1}$ is a minimal element and $x_{j_\ell}$ is a maximal element of the poset. The *rank* of $P$ is $\sharp(C) - 1$, where $C$ is a chain with maximum length of $P$.

The *order polytope* of $P$ is the convex polytope $\mathcal{O}(P) \subset \mathbb{R}^d$ which consists of those $(a_1, \ldots a_d) \in \mathbb{R}^d$ such that $0 \leqslant a_i \leqslant 1$ for every $1 \leqslant i \leqslant d$ together with

$$a_i \geqslant a_j$$

if $x_i \leqslant x_j$ in $P$.

The *chain polytope* of $P$ is the convex polytope $\mathscr{C}(P) \subset \mathbb{R}^d$ which consists of those $(a_1, \ldots, a_d) \in \mathbb{R}^d$ such that $a_i \geq 0$ for every $1 \leq i \leq d$ together with

$$a_{i_1} + a_{i_2} + \cdots + a_{i_k} \leq 1$$

for every maximal chain $x_{i_1} < x_{i_2} < \cdots < x_{i_k}$ of $P$.

One has $\dim \mathscr{O}(P) = \dim \mathscr{C}(P) = d$. The vertices of $\mathscr{O}(P)$ is those $\rho(I)$ for which $I$ is a poset ideal of $P$ ([1] (Corollary1.3)) and the vertices of $\mathscr{C}(P)$ is those $\rho(A)$ for which $A$ is an antichain of $P$ ([1] (Theorem2.2)). It then follows that the number of vertices of $\mathscr{O}(P)$ is equal to that of $\mathscr{C}(P)$. Moreover, the volume of $\mathscr{O}(P)$ and that of $\mathscr{C}(P)$ are equal to $e(P)/d!$, where $e(P)$ is the number of linear extensions of $P$ ([1] (Corollary4.2)). It also follows from [1] that the facets of $\mathscr{O}(P)$ are the following:

- $x_i = 0$, where $x_i \in P$ is maximal;
- $x_j = 1$, where $x_j \in P$ is minimal;
- $x_i = x_j$, where $x_j$ covers $x_i$,

and that the facets of $\mathscr{C}(P)$ are the following:

- $x_i = 0$, for all $x_i \in P$;
- $x_{i_1} + \cdots + x_{i_k} = 1$, where $x_{i_1} < \cdots < x_{i_k}$ is a maximal chain of $P$.

In [4] a characterization of edges of $\mathscr{O}(P)$ and those of $\mathscr{C}(P)$ is obtained. Recall that a subposet $Q$ of finite poset $P$ is said to be *connected* in $P$ if, for each $x$ and $y$ belonging to $Q$, there exists a sequence $x = x_0, x_1, \ldots, x_s = y$ with each $x_i \in Q$ for which $x_{i-1}$ and $x_i$ are comparable in $P$ for each $1 \leq i \leq s$.

**Lemma 1** ([4] (Lemma 4, Lemma 5)). *Let $P$ be a finite poset.*

1. *Let $I$ and $J$ be poset ideals of $P$ with $I \neq J$. Then the convex hull of $\{\rho(I), \rho(J)\}$ forms an edge of $\mathscr{O}(P)$ if and only if $I \subset J$ and $J \setminus I$ is connected in $P$.*
2. *Let $A$ and $B$ be antichains of $P$ with $A \neq B$. Then the convex hull of $\{\rho(A), \rho(B)\}$ forms an edge of $\mathscr{C}(P)$ if and only if $(A \setminus B) \cup (B \setminus A)$ is connected in $P$.*

### 3. Faces of 2-Dimensional Simplex

Using Lemma 1, we show the following description of faces of 2-dimensional simplex.

**Theorem 1.** *Let $P$ be a finite poset. Let $I$, $J$, and $K$ be pairwise distinct poset ideals of $P$. Then the convex hull of $\{\rho(I), \rho(J), \rho(K)\}$ forms a 2-face of $\mathscr{O}(P)$ if and only if $I \subset J \subset K$ and $K \setminus I$ is connected in $P$.*

**Proof.** ("Only if") If the convex hull of $\{\rho(I), \rho(J), \rho(K)\}$ forms a 2-face of $\mathscr{O}(P)$, then the convex hulls of $\{\rho(I), \rho(J)\}$, $\{\rho(J), \rho(K)\}$, and $\{\rho(I), \rho(K)\}$ form edges of $\mathscr{O}(P)$. It then follows from Lemma 1 that $I \subset J \subset K$ and $K \setminus I$ is connected in $P$.

("If") Suppose that the convex hull of $\{\rho(I), \rho(J), \rho(K)\}$ has dimension 1. Then there exists a line passing through the lattice points $\rho(I), \rho(J)$, and $\rho(K)$. Hence $\rho(I), \rho(J)$, and $\rho(K)$ cannot be vertices of $\mathscr{O}(P)$. Thus the convex hull of $\{\rho(I), \rho(J), \rho(K)\}$ has dimension 2.

Let $P = \{x_1, \ldots, x_d\}$. If there exists a maximal element $x_i$ of $P$ not belonging to $I \cup J \cup K$, then the convex hull of $\{\rho(I), \rho(J), \rho(K)\}$ lies in the facet $x_i = 0$. If there exists a minimal element $x_j$ of $P$ belonging to $I \cap J \cap K$, then the convex hull of $\{\rho(I), \rho(J), \rho(K)\}$ lies in the facet $x_j = 1$. Hence, working with induction on $d(\geq 2)$, we may assume that $I \cup J \cup K = P$ and $I \cap J \cap K = \emptyset$. Suppose that $\emptyset = I \subset J \subset K = P$ and $K \setminus I = P$ is connected.

**Case 1.** $\sharp(J) = 1$.

Let $J = \{x_i\}$ and $P' = P\backslash\{x_i\}$. Then $P'$ is a connected poset. Let $x_{i_1},\ldots,x_{i_q}$ be the maximal elements of $P$ and $\mathcal{A}_{ij} = \{y \in P' \mid y < x_{i_j}\}$, where $1 \leq j \leq q$. Then we write

$$b_k = \begin{cases} \sharp(\{i_j \mid x_k \in \mathcal{A}_{ij}\}) & \text{if } k \notin \{i_1,\ldots,i_q,i\} \\ 0 & \text{if } k = i \\ -\sharp(\mathcal{A}_{ij}) & \text{if } k \in \{i_1,\ldots,i_q\} \end{cases}.$$

We then claim that the hyperplane $\mathcal{H}$ of $\mathbb{R}^d$ defined by the equation $h(\mathbf{x}) = \sum_{k=1}^d b_k x_k = 0$ is a supporting hyperplane of $\mathcal{O}(P)$ and that $\mathcal{H} \cap \mathcal{O}(P)$ coincides with the convex hull of $\{\rho(\varnothing),\rho(J),\rho(P)\}$. Clearly $h(\rho(\varnothing)) = h(\rho(P)) = 0$ and $h(\rho(J)) = b_i = 0$. Let $I$ be a poset ideal of $P$ with $I \neq \varnothing$, $I \neq P$ and $I \neq J$. We have to prove that $h(\rho(I)) > 0$. To simplify the notation, suppose that $I \cap \{x_{i_1},\ldots,x_{i_q}\} = \{x_{i_1},\ldots,x_{i_r}\}$, where $0 \leq r < q$. If $r = 0$, then $h(\rho(J)) > 0$. Let $1 \leq r < q$, $I' = I\backslash\{x_i\}$, and $K = \bigcup_{j=1}^r (\mathcal{A}_{i_j} \cup \{x_{i_j}\})$. Then $I'$ and $K$ are poset ideals of $P$ and $h(\rho(K)) \leq h(\rho(I')) = h(\rho(I))$. We claim $h(\rho(K)) > 0$. One has $h(\rho(K)) \geq 0$. Moreover, $h(\rho(K)) = 0$ if and only if no $z \in K$ belongs to $\mathcal{A}_{i_{r+1}} \cup \cdots \cup \mathcal{A}_{i_q}$. Now, since $P'$ is connected, it follows that there exists $z \in K$ with $z \in \mathcal{A}_{i_{r+1}} \cup \cdots \cup \mathcal{A}_{i_q}$. Hence $h(\rho(K)) > 0$. Thus $h(\rho(I)) > 0$.

**Case 2.** $\sharp(J) = d - 1$.

Let $P\backslash J = \{x_i\}$ and $P' = P\backslash\{x_i\}$. Then $P'$ is a connected poset. Thus we can show the existence of a supporting hyperplane of $\mathcal{O}(P)$ which contains the convex hull of $\{\rho(\varnothing),\rho(J),\rho(P)\}$ by the same argument in Case 1.

**Case 3.** $2 \leq \sharp(J) \leq d - 2$.

To simplify the notation, suppose that $J = \{x_1,\ldots,x_\ell\}$. Then $P\backslash J = \{x_{\ell+1},\ldots,x_d\}$. Since $J$ and $P\backslash J$ are subposets of $P$, these posets are connected. Let $x_{i_1},\ldots,x_{i_q}$ be the maximal elements of $J$ and $x_{i_{q+1}},\ldots,x_{i_{q+r}}$ the maximal elements of $P\backslash J$. Then we write

$$\mathcal{A}_{ij} = \begin{cases} \{y \in J \mid y < x_{i_j}\} & \text{if } 1 \leq j \leq q \\ \{y \in P\backslash J \mid y < x_{i_j}\} & \text{if } q+1 \leq j \leq r \end{cases}$$

and

$$b_k = \begin{cases} \sharp(\{i_j \mid x_i \in \mathcal{A}_{ij}\}) & \text{if } k \notin \{i_1,\ldots,i_q,i_{q+1},\ldots,i_{q+r}\} \\ -\sharp(\mathcal{A}_{ij}) & \text{if } k \in \{i_1,\ldots,i_q,i_{q+1},\ldots,i_{q+r}\} \end{cases}.$$

We then claim that the hyperplane $\mathcal{H}$ of $\mathbb{R}^d$ defined by the equation $h(\mathbf{x}) = \sum_{k=1}^d b_k x_k = 0$ is a supporting hyperplane of $\mathcal{O}(P)$ and $\mathcal{H} \cap \mathcal{O}(P)$ coincides with the convex hull of $\{\rho(\varnothing),\rho(J),\rho(P)\}$. Clearly $h(\rho(\varnothing)) = h(\rho(J)) = h(\rho(P\backslash J)) = 0$, then $h(\rho(P)) = h(\rho(J)) + h(\rho(P\backslash J)) = 0$. Let $I$ be a poset ideal of $P$ with $I \neq \varnothing$, $I \neq P$ and $I \neq J$. What we must prove is $h(\rho(I)) > 0$.

If $I \subset J$, then $I$ is a poset ideal of $J$. To simplify the notation, suppose that $I \cap \{x_{i_1},\ldots,x_{i_q}\} = \{x_{i_1},\ldots,x_{i_s}\}$, where $0 \leq s < q$. If $s = 0$, then $h(\rho(I)) > 0$. Let $1 \leq s < q$, $K = \bigcup_{j=1}^s (\mathcal{A}_{i_j} \cup \{x_{i_j}\})$. Then $K$ is a poset ideal of $J$ and $h(\rho(K)) \leq h(\rho(I))$. Thus we can show $h(\rho(K)) > 0$ by the same argument in Case 1 (Replace $r$ with $s$ and $P'$ with $J$).

If $J \subset I$, then $I\backslash J$ is a poset ideal of $P\backslash J$. To simplify the notation, suppose that $(I\backslash J) \cap \{x_{i_{q+1}},\ldots,x_{i_{q+r}}\} = \{x_{i_{q+1}},\ldots,x_{i_{q+t}}\}$, where $0 \leq t < r$. If $t = 0$, then $h(\rho(I)) = h(\rho(J)) + h(\rho(I\backslash J)) = h(\rho(I\backslash J)) > 0$. Let $1 \leq t < r$, $K = \bigcup_{j=q+1}^{q+t}(\mathcal{A}_{i_j} \cup \{x_{i_j}\})$. Then $K$ is a poset ideal of $P\backslash J$ and $h(\rho(K)) \leq h(\rho(I\backslash J)) = h(\rho(I))$. Thus we can show $h(\rho(K)) > 0$ by the same argument in Case 1 (Replace $r$ with $q+t$, $q$ with $q+r$ and $P'$ with $P\backslash J$). Consequently, $h(\rho(I)) > 0$, as desired. □

Let $A \triangle B$ denote the symmetric difference of the sets $A$ and $B$, that is $A \triangle B = (A\backslash B) \cup (B\backslash A)$.

**Theorem 2.** *Let $P$ be a finite poset. Let $A$, $B$, and $C$ be pairwise distinct antichains of $P$. Then the convex hull of $\{\rho(A),\rho(B),\rho(C)\}$ forms a 2-face of $\mathcal{C}(P)$ if and only if $A \triangle B$, $B \triangle C$ and $C \triangle A$ are connected in $P$.*

**Proof.** ("Only if") If the convex hull of $\{\rho(A), \rho(B), \rho(C)\}$ forms a 2-face of $\mathscr{C}(P)$, then the convex hulls of $\{\rho(A), \rho(B)\}$, $\{\rho(B), \rho(C)\}$, and $\{\rho(A), \rho(C)\}$ form edges of $\mathscr{C}(P)$. It then follows from Lemma 1 that $A \triangle B$, $B \triangle C$ and $C \triangle A$ are connected in $P$.

("If") Suppose that the convex hull of $\{\rho(A), \rho(B), \rho(C)\}$ has dimension 1. Then there exists a line passing through the lattice points $\rho(A)$, $\rho(B)$, and $\rho(C)$. Hence $\rho(A)$, $\rho(B)$, and $\rho(C)$ cannot be vertices of $\mathscr{C}(P)$. Thus the convex hull of $\{\rho(A), \rho(B), \rho(C)\}$ has dimension 2.

Let $P = \{x_1, \ldots, x_d\}$. If $A \cup B \cup C \neq P$ and $x_i \notin A \cup B \cup C$, then the convex hull of $\{\rho(A), \rho(B), \rho(C)\}$ lies in the facet $x_i = 0$. Furthermore, if $A \cup B \cup C = P$ and $A \cap B \cap C \neq \emptyset$, then $x_j \in A \cap B \cap C$ is isolated in $P$ and $x_j$ itself is a maximal chain of $P$. Thus the convex hull of $\{\rho(A), \rho(B), \rho(C)\}$ lies in the facet $x_j = 1$. Hence, working with induction on $d (\geq 2)$, we may assume that $A \cup B \cup C = P$ and $A \cap B \cap C = \emptyset$. As stated in the proof of [3] ([Theorem 2.1]), if $A \triangle B$ is connected in $P$, then $A$ and $B$ satisfy either (i) $B \subset A$ or (ii) $y < x$ whenever $x \in A$ and $y \in B$ are comparable. Hence, we consider the following three cases:

(a) If $B \subset A$, then $A \triangle B = A \setminus B$ is connected in $P$, and thus $\sharp(A \setminus B) = 1$. Let $A \setminus B = \{x_k\}$. If $C \cap A \neq \emptyset$, then $C \cap A = \{x_k\}$, since $A \cap B \cap C = C \cap B = \emptyset$. Namely $x_k$ is isolated in $P$. Hence $B \triangle C = B \cup C = A \cup B \cup C = P$ cannot be connected. Thus $C \cap A = \emptyset$. In this case, we may assume $z < x$ if $x \in A$ and $z \in C$ are comparable. Furthermore, $P$ has rank 1.

(b) If $B \not\subset A$ and $B \cap A \neq \emptyset$, then we may assume $y < x$ if $x \in A$ and $y \in B$ are comparable. If $C \subset B$ with $C \cap A \cap B = \emptyset$, then as stated in (a), $C \triangle A$ cannot be connected. Since $C \not\subset B$, we may assume $z < y$ if $y \in B$ and $z \in C$ are comparable. If $C \cap B \neq \emptyset$, then $C \cap A = \emptyset$ and $P$ has rank 1 or 2. Similarly, if $C \cap B = \emptyset$, then $C \cap A = \emptyset$ and $P$ has rank 2.

(c) Let $B \not\subset A$ and $B \cap A = \emptyset$. We may assume that if $x \in A$ and $y \in B$ are comparable, then $y < x$. If $C \subset B$, then we regard this case as equivalent to (a). Let $C \not\subset B$. We may assume $z < y$ if $y \in B$ and $z \in C$ are comparable. Moreover, if $C \cap B \neq \emptyset$, then we regard this case as equivalent to (b). If $C \cap B = \emptyset$, then $C \cap A = \emptyset$ and $P$ has rank 2.

Consequently, there are five cases as regards antichains for $\mathscr{C}(P)$.

**Case 1.** $B \subset A$, $C \cap A = \emptyset$, and $C \cap B = \emptyset$.

For each $x_i \in B$ we write $b_i$ for the number of elements $z \in C$ with $z < x_i$. For each $x_j \in C$ we write $c_j$ for the number of elements $y \in B$ with $x_j < y$. Let $a_k = 0$ for $A \setminus B = \{x_k\}$. Clearly $\sum_{x_i \in B} b_i = \sum_{x_j \in C} c_j = q$, where $q$ is the number of pairs $(y, z)$ with $y \in B$, $z \in C$ and $z < y$. Let $h(\mathbf{x}) = \sum_{x_i \in B} b_i x_i + \sum_{x_j \in C} c_j x_j + a_k x_k$ and let $\mathscr{H}$ be the hyperplane of $\mathbb{R}^d$ defined by $h(\mathbf{x}) = q$. Then $h(\rho(A)) = h(\rho(B)) = h(\rho(C)) = q$. We claim that, for any antichain $D$ of $P$ with $D \neq A$, $D \neq B$, and $D \neq C$, one has $h(\rho(D)) < q$. Let $D = B_1 \cup C_1$ or $D = \{x_k\} \cup C_1$ with $B_1 \subsetneq B$ and $C_1 \subsetneq C$. Suppose $D = B_1 \cup C_1$. Since $B \triangle C$ is connected and since $D$ is an antichain of $P$, it follows that $\sum_{x_i \in B_1} b_i + \sum_{x_j \in C_1} c_j < q$. Thus $h(\rho(D)) < q$. Suppose that $D = \{x_k\} \cup C_1$. It follows that $\sum_{x_j \in C_1} c_j + a_k = \sum_{x_j \in C_1} c_j < \sum_{x_j \in C} c_j = q$. Thus $h(\rho(D)) < q$.

**Case 2.** $B \not\subset A$, $B \cap A \neq \emptyset$, $C \not\subset B$, $C \cap B \neq \emptyset$, $C \cap A = \emptyset$, and $P$ has rank 1.

We define four numbers as follows:

$$\alpha_i = \sharp(\{y \in B \setminus A \mid y < x_i, \ x_i \in A \setminus B\});$$
$$\gamma_j = \sharp(\{x \in A \setminus B \mid x_j < x, \ x_j \in B \setminus A\});$$
$$\alpha_k = \sharp(\{z \in C \setminus B \mid z < x_k, \ x_k \in B \setminus C\});$$
$$\gamma_\ell = \sharp(\{y \in B \setminus C \mid x_\ell < y, \ x_\ell \in C \setminus B\}).$$

Since $P$ has rank 1, $B \subset A \cup C = P$. It follows that $A = (A\backslash B) \cup (B\backslash C)$, $C = (B\backslash A) \cup (C\backslash B)$. Then

$$\sum_{x_s \in A} \alpha_s = \sum_{x_i \in A\backslash B} \alpha_i + \sum_{x_k \in B\backslash C} \alpha_k = q;$$

$$\sum_{x_j \in B\backslash A} \gamma_j + \sum_{x_k \in B\backslash C} \alpha_k = q;$$

$$\sum_{x_u \in C} \gamma_u = \sum_{x_j \in B\backslash A} \gamma_j + \sum_{x_\ell \in C\backslash B} \gamma_\ell = q,$$

where $q_1$ is the number of pairs $(x,y)$ with $x \in A\backslash B$, $y \in B\backslash A$ and $y < x$, $q_2$ is the number of pairs $(y,z)$ with $y \in B\backslash C$, $z \in C\backslash B$ and $z < y$, and $q = q_1 + q_2$. Let

$$h(\mathbf{x}) = \sum_{x_s \in A} \alpha_s x_s + \sum_{x_u \in C} \gamma_u x_u$$

$$= \sum_{x_i \in A\backslash B} \alpha_i x_i + \left( \sum_{x_j \in B\backslash A} \gamma_j x_j + \sum_{x_k \in B\backslash C} \alpha_k x_k \right) + \sum_{x_\ell \in C\backslash B} \gamma_\ell x_\ell$$

and $\mathcal{H}$ the hyperplane of $\mathbb{R}^d$ defined by $h(\mathbf{x}) = q$. Then $h(\rho(A)) = h(\rho(B)) = h(\rho(C)) = q$. We claim that, for any antichain $D$ of $P$ with $D \neq A$, $D \neq B$ and $D \neq C$, one has $h(\rho(D)) < q$. Let $D = D_1 \cup D_2$ with $D_1$ is an antichain of $A \triangle B$ and $D_2$ is an antichain of $B \triangle C$. Since $A \triangle B$, $B \triangle C$ are connected, it follows that $h(\rho(D_1)) < q_1$ and $h(\rho(D_2)) < q_2$. Thus $h(\rho(D)) = h(\rho(D_1)) + h(\rho(D_2)) < q_1 + q_2 = q$.

**Case 3.** $B \not\subset A$, $B \cap A \neq \emptyset$, $C \not\subset B$, $C \cap B \neq \emptyset$, $C \cap A = \emptyset$, and $P$ has rank 2.

For each $x_i \in P$ we write $c(i)$ for the number of maximal chains, which contain $x_i$. Let $q$ be the number of maximal chains in $P$. Since each $x_i \in A$ is maximal element and each $x_k \in C$ is minimal element, $\sum_{x_i \in A} c(i) = \sum_{x_k \in C} c(k) = q$. Then

$$\sum_{x_j \in B} c(j) = \sum_{x_s \in B \cap A} c(s) + \sum_{x_t \in B \cap C} c(t) + \sum_{x_u \in B \backslash (A \cup C)} c(u)$$

$$= \sum_{x_s \in B \cap A} c(s) + \sum_{x_t \in B \cap C} c(t) + \left( \sum_{x_v \in A\backslash B} c(v) - \sum_{x_t \in B \cap C} c(t) \right)$$

$$= \sum_{x_i \in A} c(i) = q.$$

Let $h(\mathbf{x}) = \sum_{x_i \in P} c(i) x_i$ and $\mathcal{H}$ the hyperplane of $\mathbb{R}^d$ defined by $h(\mathbf{x}) = q$. Then $h(\rho(A)) = h(\rho(B)) = h(\rho(C)) = q$. We claim that, for any antichain $D$ of $P$ with $D \neq A$, $D \neq B$ and $D \neq C$, one has $h(\rho(D)) < q$. $D = A_1 \cup B_1 \cup C_1$ with $A_1 \subset A\backslash B$, $B_1 \subsetneq B$, and $C_1 \subsetneq C\backslash B$. Now, we define two subsets of $B$:

$$B_2 = \{x_j \in B \mid x_j < x_i, x_i \in A_1\};$$
$$B_3 = \{x_j \in B \mid x_k < x_j, x_k \in C_1\}.$$

Then $B_1 \cap B_2 = B_1 \cap B_3 = B_2 \cap B_3 = \emptyset$ and $B_1 \cup B_2 \cup B_3 \subset B_3$. Let $\sum_{x_i \in A} c(i) = q_1$, $\sum_{x_j \in B_1} c(j) = q_2$, $\sum_{x_k \in C_1} c(k) = q_3$, $\sum_{x_j \in B_2} c(j) = q'_1$, and $\sum_{x_j \in B_3} c(j) = q'_3$. Since $A \triangle B$, $B \triangle C$ are connected, it follows that $q_1 < q'_1$ and $q_3 < q'_3$. Hence

$$h(\rho(D)) = \sum_{x_i \in A_1} c(i) + \sum_{x_j \in B_1} c(j) + \sum_{x_k \in C_1} c(k)$$

$$= q_1 + q_2 + q_3 < q'_1 + q_2 + q'_3$$

$$= \sum_{x_j \in B_2} c(j) + \sum_{x_j \in B_1} c(j) + \sum_{x_j \in B_3} c(j) \leq \sum_{x_j \in B} c(j) = q.$$

Thus $h(\rho(D)) < q$.

**Case 4.** $B \not\subseteq A$, $B \cap A \neq \emptyset$, $C \cap B = \emptyset$, and $C \cap A = \emptyset$.

Since $P$ has rank 2, we can show $h(\rho(D)) < q$ by the same argument in Case 3 (Suppose $C \cap B = \emptyset$).

**Case 5.** $B \not\subseteq A$, $B \cap A = \emptyset$, $C \cap B = \emptyset$ and $C \cap A = \emptyset$.

Since $P$ has rank 2, we can show $h(\rho(D)) < q$ by the same argument in Case 3 (Suppose $B \cap A = C \cap B = \emptyset$).

In conclusion, each $\mathcal{H}$ is a supporting hyperplane of $\mathscr{C}(P)$ and $\mathcal{H} \cap \mathscr{C}(P)$ coincides with the convex hull of $\{\rho(A), \rho(B), \rho(C)\}$, as desired. □

**Corollary 1.** *Triangles in 1-skeleton of $\mathcal{O}(P)$ or $\mathscr{C}(P)$ are in one-to-one correspondence with faces of 2-dimensional simplex of each polytope.*

**Funding:** This research received no external funding.

**Conflicts of Interest:** The author declares no conflict of interest.

## References

1. Stanley, R. Two poset polytopes. *Discrete Comput. Geom.* **1986**, *1*, 9–23. [CrossRef]
2. Hibi, T.; Li, N. Unimodular equivalence of order and chain polytopes. *Math. Scand.* **2016**, *118*, 5–12. [CrossRef]
3. Hibi, T.; Li, N.; Sahara, Y.; Shikama, A. The numbers of edges of the order polytope and the chain polytope of a finite partially ordered set. *Discret. Math.* **2017**, *340*, 991–994. [CrossRef]
4. Hibi, T.; Li, N. Cutting convex polytopes by hyperplanes. *Mathematics* **2019**, *7*, 381. [CrossRef]

© 2019 by the authors. Licensee MDPI, Basel, Switzerland. This article is an open access article distributed under the terms and conditions of the Creative Commons Attribution (CC BY) license (http://creativecommons.org/licenses/by/4.0/).

Article
# On the Stanley Depth of Powers of Monomial Ideals

S. A. Seyed Fakhari [1,2]

[1] School of Mathematics, Statistics and Computer Science, College of Science, University of Tehran, Tehran, Iran; aminfakhari@ut.ac.ir
[2] Institute of Mathematics, Vietnam Academy of Science and Technology, 18 Hoang Quoc Viet, Hanoi, Vietnam

Received: 7 June 2019; Accepted: 4 July 2019; Published: 8 July 2019

**Abstract:** In 1982, Stanley predicted a combinatorial upper bound for the depth of any finitely generated multigraded module over a polynomial ring. The predicted invariant is now called the Stanley depth. Duval et al. found a counterexample for Stanley's conjecture, and their counterexample is a quotient of squarefree monomial ideals. On the other hand, there is evidence showing that Stanley's inequality can be true for high powers of monomial ideals. In this survey article, we collect the recent results in this direction. More precisely, we investigate the Stanley depth of powers, integral closure of powers, and symbolic powers of monomial ideals.

**Keywords:** complete intersection; cover ideal; depth; edge ideal; integral closure; polymatroidal ideal; Stanley depth; Stanley's inequality; symbolic power

## 1. Introduction

Let $\mathbb{K}$ be a field, and set $S = \mathbb{K}[x_1, \ldots, x_n]$. Assume that $M$ is a finitely generated, $\mathbb{Z}^n$-graded $S$-module. For any homogeneous element $u \in M$ and any $Z \subseteq \{x_1, \ldots, x_n\}$, the $\mathbb{K}$-vector space $u\mathbb{K}[Z]$ is called a Stanley space of dimension $|Z|$. A Stanley decomposition of $M$ is a decomposition of $M$ as a finite direct sum of Stanley spaces. The minimum dimension of a Stanley space in a Stanley decomposition $\mathcal{D}$ is called the Stanley depth of $\mathcal{D}$, and is denoted by $\mathrm{sdepth}(\mathcal{D})$. The Stanley depth of $M$ is defined

$$\mathrm{sdepth}(M) := \max\left\{\mathrm{sdepth}(\mathcal{D}) \mid \mathcal{D} \text{ is a Stanley decomposition of } M\right\}.$$

As a convention, we set $\mathrm{sdepth}(M) = \infty$ when $M$ is the zero module. For an introduction to Stanley depth, we refer the reader to [1].

**Example 1.** *Consider the ideal $I = \langle x_1 x_2^2, x_1^2 x_2 \rangle$ in the polynomial ring $S = \mathbb{K}[x_1, x_2]$. Then,*

$$\mathcal{D}_1: \ I = x_1 x_2^2 \mathbb{K}[x_2] \oplus x_1^2 x_2 \mathbb{K}[x_1, x_2]$$

*is a Stanley decomposition for $I$, with $\mathrm{sdepth}(\mathcal{D}_1) = 1$. One can also write other Stanley decompositions for $I$. For example,*

$$\mathcal{D}_2: \ I = x_1^2 x_2 \mathbb{K} \oplus x_1^3 x_2 \mathbb{K}[x_1] \oplus x_1 x_2^2 \mathbb{K}[x_2] \oplus x_1^2 x_2^2 \mathbb{K}[x_1, x_2],$$

*and*

$$\mathcal{D}_3: \ I = x_1^2 x_2 \mathbb{K} \oplus x_1^3 x_2 \mathbb{K} \oplus x_1^4 x_2 \mathbb{K}[x_1] \oplus x_1 x_2^2 \mathbb{K}[x_2] \oplus x_1^2 x_2^2 \mathbb{K}[x_1, x_2].$$

*It is clear that $\mathrm{sdepth}(\mathcal{D}_2) = \mathrm{sdepth}(\mathcal{D}_3) = 0$. It follows from the definition of Stanley depth that $\mathrm{sdepth}(I) \geq 1$, and it can be easily verified that the equality indeed holds—that is, $\mathrm{sdepth}(I) = 1$.*

We say that a $\mathbb{Z}^n$-graded $S$-module $M$ satisfies *Stanley's inequality* if

$$\operatorname{depth}(M) \leq \operatorname{sdepth}(M).$$

In fact, Stanley [2] conjectured that the above inequality holds for every finitely generated, $\mathbb{Z}^n$-graded $S$-module. Ichim, Katthän, and Moyano-Fernández [3] showed that in order to prove Stanley's conjecture for monomial ideals, it is enough to consider squarefree monomial ideals. However, Stanley's conjecture has been disproved by Duval, Goeckner, Klivans, and Martin [4]. In fact, they constructed a non-partitionable Cohen-Macaulay simplicial complex, and then, using a result of Herzog, Soleyman Jahan, and Yassemi ([5] Corollary 4.5), deduced that the Stanley-Reisner ring of this simplicial complex did not satisfy Stanley's inequality. However, it is still interesting to find new classes of modules which satisfy Stanley's inequality. Of particular interest is the validity of Stanley's inequality for high powers of monomial ideals. In this survey article, we review the recent developments in this regard. In 2013, Herzog [6] published his nice survey on Stanley depth. In fact, we complement his survey by collecting the results obtained since then, with a focus on powers of monomial ideals.

## 2. Ordinary Powers

In this section, we consider the ordinary powers of monomial ideals. As we explained in the introduction, it is natural to ask whether the high powers of monomial ideals satisfy Stanley's inequality. In fact, this question was posed in [7].

**Question 1** ([7], Question 1.1). *Let $I$ be a monomial ideal. Is it true that $I^k$ and $S/I^k$ satisfy Stanley's inequality for every integer, $k \gg 0$?*

In the following subsections we will see that Question 1 has a positive answer when $I$ belongs to interesting classes of monomial ideals.

### 2.1. Maximal Ideal and Complete Intersections

Let $\mathfrak{m} = (x_1, \ldots, x_n)$ denote the maximal graded ideal of $S$. It is clear that for every integer $k \geq 1$, $\operatorname{depth}(S/\mathfrak{m}^k) = 0$. Hence, $S/\mathfrak{m}^k$ satisfies Stanley's inequality for any $k \geq 1$. Indeed, since $S/\mathfrak{m}^k$ is an Artinian ring, we also have $\operatorname{sdepth}(S/\mathfrak{m}^k) = 0$ for every integer, $k \geq 1$. On the other hand, $\operatorname{depth}(\mathfrak{m}^k) = 1$ and by ([6] Corollary 24), we know that the Stanley depth of any monomial ideal is at least one. This implies that $\mathfrak{m}^k$ satisfies Stanley's inequality for every integer, $k \geq 1$. However, computing the exact value of the Stanley depth of $\mathfrak{m}^k$ is not easy. In 2010, Biró, Howard, Keller, Trotter, and Young [8] proved that $\operatorname{sdepth}(\mathfrak{m}) = \lceil n/2 \rceil$. Cimpoeaş [9] determined an upper bound for the Stanley depth of powers of $\mathfrak{m}$. More precisely, he proved the following results:

**Theorem 1** ([9], Theorem 2.2). *For every integer $k \geq 1$, we have*

$$\operatorname{sdepth}(\mathfrak{m}^k) \leq \left\lceil \frac{n}{k+1} \right\rceil.$$

*In particular, for every integer $k \geq n-1$, we have $\operatorname{sdepth}(\mathfrak{m}^k) = 1$.*

Cimpoeaş [9] also conjectured that the inequality obtained in the above theorem was indeed an equality—that is,

$$\operatorname{sdepth}(\mathfrak{m}^k) = \left\lceil \frac{n}{k+1} \right\rceil,$$

for every $k \geq 1$.

In 2018, Cimpoeaş [10] extended Theorem 1 by determining the bounds for the Stanley depth of complete intersection monomial ideals.

**Theorem 2** ([10], Proposition 2.14 and Theorem 2.15). *Let I be a complete intersection monomial ideal which is minimally generated by t monomials.*

(i) *For every integer $k \geq 1$, we have*

$$n - t + 1 \leq \mathrm{sdepth}(I^k) \leq n - t + \left\lceil \frac{t}{k+1} \right\rceil.$$

*In particular, if $k \geq t - 1$, then $\mathrm{sdepth}(I^k) = n - t + 1$.*

(ii) *For every integer $k \geq 1$, we have*

$$\mathrm{sdepth}(S/I^k) = \mathrm{sdepth}(I^k/I^{k+1}) = \dim(S/I) = n - t.$$

As an immediate consequence of Theorem 2, we conclude that for any complete intersection monomial ideal and every integer $k \geq 1$, the modules $I^k, S/I^k$, and $I^k/I^{k+1}$ satisfy Stanley's inequality. In particular, Question 1 has a positive answer in this case.

*2.2. Polymatroidal Ideals*

We begin this subsection by recalling the definition of polymatroidal ideals. We mention that for every monomial ideal $I$, we denote its minimal set of monomial generators by $G(I)$.

**Definition 1.** *A monomial ideal I is called polymatroidal if it is generated in a single degree, and moreover, for every pair of monomials $u = x_1^{a_1}, \ldots, x_n^{a_n}$ and $v = x_1^{b_1}, \ldots, x_n^{b_n}$ belonging to $G(I)$, and for every $i$ with $a_i > b_i$, one has $j$ with $a_j < b_j$, such that $x_j(u/x_i) \in G(I)$.*

We next define the class of weakly polymatroidal ideals, which is a generalization of the class of polymatroidal ideals.

**Definition 2** ([11], Definition 1.1). *A monomial ideal I is called weakly polymatroidal if, for every pair of monomials, $u = x_1^{a_1} \ldots x_n^{a_n}$ and $v = x_1^{b_1} \ldots x_n^{b_n}$ in $G(I)$ such that $a_1 = b_1, \ldots, a_{t-1} = b_{t-1}$ and $a_t > b_t$ for some $t$, there exists $j > t$ such that $x_t(v/x_j) \in I$.*

It is obvious that any polymatroidal ideal is weakly polymatroidal.

Let $I$ be a weakly polymatroidal ideal. In ([12] Theorem 2.4), we proved that $S/I$ satisfies Stanley's inequality. We also know from ([13] Theorem 12.6.3) that every power of a polymatroidal ideal is again a polymatroidal ideal. As a consequence, for any polymatroidal ideal $I$ and any integer $k \geq 1$, the module $S/I^k$ satisfies Stanley's inequality. It is natural to ask whether $I^k$ satisfies Stanley's inequality. Before answering this question, we recall the concept of having linear quotients, introduced in [14].

**Definition 3.** *Let I be a monomial ideal, and assume that $G(I)$ is the set of minimal monomial generators of I. We say that I has linear quotients if there is a linear order $u_1 \prec u_2 \prec \ldots \prec u_m$ on $G(I)$, with the property that for every $2 \leq i \leq m$, the ideal $(u_1, \ldots, u_{i-1}) : u_i$ is generated by a subset of the variables.*

Soleyman Jahan [15] proves that Stanley's inequality holds for any monomial ideal which has linear quotients. On the other hand, by ([11] Theorem 1.3), we know that any weakly polymatroidal ideal has linear quotients. This implies that every weakly polymatroidal ideal satisfies Stanley's inequality. Since every power of a polymatroidal ideal is again a polymatroidal ideal, we deduce that for any polymatroidal ideal $I$ and any integer $k \geq 1$, the ideal $I^k$ satisfies Stanley's inequality.

By the above argument, we know that Question 1 has a positive answer for polymatroidal ideals. This result was also obtained in [16].

Let $I$ be a monomial ideal of $S$ with the Rees algebra $\mathcal{R}(I) = \bigoplus_{k=0}^{\infty} I^k$. The $\mathbb{K}$-algebra $\mathcal{R}(I)/\mathfrak{m}\mathcal{R}(I)$ is called the *fibre ring*, and its Krull dimension is called the *analytic spread* of $I$, denoted by $\ell(I)$. A classical result by Burch [17] states that

$$\min_k \operatorname{depth}(S/I^k) \leq n - \ell(I).$$

By a theorem of Brodmann [18], $\operatorname{depth}(S/I^k)$ is constant for large $k$. We call this constant value the *limit depth* of $I$, and denote it by $\lim_{k\to\infty} \operatorname{depth}(S/I^k)$. Brodmann improved Burch's inequality by showing that

$$\lim_{k\to\infty} \operatorname{depth}(S/I^k) \leq n - \ell(I).$$

We know from ([19] Corollary 3.5) that equality occurs in the above inequality if $I$ is a polymatroidal ideal. In fact, we will see in the next section that equality holds in Burch's inequality for a larger class of ideals—namely, the class of *normal* ideals.

Inspired by the limiting behavior of the depth of powers of ideals, Herzog [6] proposed the following conjecture.

**Conjecture 1** ([6], Conjecture 59). *For every monomial ideal $I$, the sequence $\{\operatorname{sdepth}(I^k)\}_{k=1}^{\infty}$ is convergent.*

This conjecture is widely open. However, by Theorem 2, it has a positive answer for complete intersections. Also, we will see in Section 4 that the assertion of this conjecture is true for any normally torsionfree, squarefree monomial ideal.

Let $I$ be a weakly polymatroidal ideal which is generated in a single degree. We know from ([16] Theorem 2.5) that $\operatorname{depth}(S/I) \geq n - \ell(I)$. Since $I$ and $S/I$ satisfy Stanley's inequality, it follows that

$$\operatorname{sdepth}(S/I) \geq n - \ell(I) \quad \text{and} \quad \operatorname{sdepth}(I) \geq n - \ell(I) + 1.$$

Restricting to the class of polymatroidal ideals, for any integer $k \geq 1$ and any polymatroidal ideal $I$, we have

$$\operatorname{sdepth}(S/I^k) \geq n - \ell(I) \quad \text{and} \quad \operatorname{sdepth}(I^k) \geq n - \ell(I) + 1.$$

Indeed, we expect that the equality holds in the above inequality for every $k \gg 0$. In other words, not only do we believe that Conjecture 1 is true for every polymatroidal ideal $I$, but we also have a prediction for the limit value of the Stanley depth of powers of $I$.

**Conjecture 2.** *Let $I$ be a polymatroidal ideal. Then,*

$$\operatorname{sdepth}(S/I^k) = n - \ell(I) \quad \text{and} \quad \operatorname{sdepth}(I^k) = n - \ell(I) + 1$$

*for any integer $k \gg 0$.*

### 2.3. Edge Ideals

Let $G$ be a finite simple graph with a vertex set $V(G) = \{x_1, \ldots, x_n\}$ and edge set $E(G)$. The *edge ideal* $I(G)$ of $G$ is defined as

$$I(G) = (x_i x_j : x_i x_j \in E(G)) \subseteq S.$$

The Stanley depth of powers of edge ideals has been studied in [20–23]. Before reviewing the main results of these papers, we mention the following result of Trung, concerning the depth of high powers of edge ideals.

**Theorem 3** ([24], Theorems 4.4 and 4.6). *Let G be a graph with n vertices and p bipartite connected components. Then, for every integer $k \geq n-1$, we have*

$$\mathrm{depth}(S/I(G)^k) = p.$$

Note that by ([25] p. 50), for every graph $G$ with $n$ vertices and $p$ bipartite connected components, we have $\ell(I(G)) = n - p$. Thus, Theorem 3, essentially says that

$$\lim_{k \to \infty} \mathrm{depth}(S/I(G)^k) = n - \ell(I(G)),$$

that is, equality occurs in Burch's inequality.

Pournaki, Yassemi, and the author [22] studied the Stanley depth of $S/I(G)^k$, where $G$ is a forest (i.e., a graph with no cycle). They proved that for every forest with $p$ connected components and any integer $k \geq 1$, we have

$$\mathrm{sdepth}(S/I(G)^k) \geq p.$$

This, together with Theorem 3, implies that for any forest $G$ with $n$ vertices, the module $S/I(G)^k$ satisfies Stanley's inequality for any integer $k \geq n-1$. This result was then extended in [23], to any arbitrary graph, as follows.

**Theorem 4** ([23], Theorem 2.3 and Corollary 2.5). *Let G be a graph with n vertices and p bipartite connected components. Then, for every integer $k \geq 1$, we have $\mathrm{sdepth}(S/I(G)^k) \geq p$. In particular, $S/I(G)^k$ satisfies Stanley's inequality for any integer $k \geq n-1$.*

We know from the above theorem that for any graph $G$, the module $S/I(G)^k$ satisfies Stanley's inequality for $k \gg 0$. However, how about $I(G)^k$? By Theorem 3, in order to prove Stanley's inequality for high powers of $I(G)$, we need to prove $\mathrm{sdepth}(I(G)^k) \geq p+1$ for every integer $k \gg 0$. We do not know whether this inequality holds for any arbitrary graph. However, we have a partial result, as follows. We recall that for any graph $G$ and every subset $U \subset V(G)$, the graph $G \setminus U$ has the vertex set $V(G \setminus U) = V(G) \setminus U$ and edge set $E(G \setminus U) = \{e \in E(G) \mid e \cap U = \emptyset\}$.

**Theorem 5** ([23], Theorem 3.1). *Let G be a graph, and assume that H is a connected component of G with at least one edge. Suppose that h is the number of bipartite connected components of $G \setminus V(H)$. Then, for every integer $k \geq 1$, we have*

$$\mathrm{sdepth}(I(G)^k) \geq \min_{1 \leq l \leq k} \{\mathrm{sdepth}_{S'}(I(H)^l)\} + h,$$

*where $S' = \mathbb{K}[x_i \mid x_i \in V(H)]$.*

Assume that $G$ has a non-bipartite connected component, and call it $H$. Then, by ([6] Corollary 24), for every integer $l \geq 1$, we have $\mathrm{sdepth}(I(H)^l) \geq 1$. Thus, it follows from Theorem 5 that in this case, $\mathrm{sdepth}(I(G)^k) \geq p+1$, where $p$ is the number of bipartite connected components of $G$ and $k \geq 1$ is an arbitrary positive integer. Assume now that $G$ is a bipartite graph. Using Theorem 5, in order to prove the inequality $\mathrm{sdepth}(I(G)^k) \geq p+1$, it is enough to prove it only for the class of connected bipartite graphs. Thus, we raise the following question.

**Question 2** ([23], Question 3.3). *Let G be a connected bipartite graph (with at least one edge) and suppose $k \geq 1$ is an integer. Is it true that $\mathrm{sdepth}(I(G)^k) \geq 2$?*

We investigated this question in [26] and proved that it has positive answer for small $k$. More precisely, we proved the following result.

**Theorem 6** ([26], Theorem 3.4). *Let G be a connected bipartite graph (with at least one edge) and let g be a positive integer. Suppose G has no cycle of length at most $g - 1$. Then, for every positive integer $k \leq g/2 + 1$, we have* $\mathrm{sdepth}(I(G)^k) \geq 2$.

Theorem 6, in particular, implies that $\mathrm{sdepth}(I(G)^k) \geq 2$, for any integer $k \geq 1$, provided that $G$ is a tree (i.e., a connected forest). Combining this result with Theorem 5 implies that if $G$ is a bipartite graph and at least one of the connected components of $G$ is a tree, then for every integer $k \geq 1$, we have $\mathrm{sdepth}(I(G)^k) \geq p + 1$, where $p$ is the number of (bipartite) connected components of $G$. All in all, we obtained the following theorem.

**Theorem 7** ([23], Corollary 3.6). *Assume that G is a graph with n vertices, such that*

(i) *G is a non-bipartite graph, or*
(ii) *at least one of the connected components of G is a tree with at least one edge.*

*Then, for every integer $k \geq n - 1$, the ideal $I(G)^k$ satisfies Stanley's inequality.*

Let $I$ be a monomial ideal. We know by ([27] Theorem 1.2) that the sequence $\{\mathrm{depth}(I^k/I^{k+1})\}_{k=1}^{\infty}$ is convergent, and moreover,

$$\lim_{k \to \infty} \mathrm{depth}(I^k/I^{k+1}) = \lim_{k \to \infty} \mathrm{depth}(S/I^k).$$

Therefore, using Theorem 3, we conclude that for any graph $G$,

$$\lim_{k \to \infty} \mathrm{depth}(I(G)^k/I(G)^{k+1}) = p,$$

where $p$ is the number of bipartite connected components of $G$. In [23], we also studied the Stanley depth of $I(G)^k/I(G)^{k+1}$ and proved that it satisfied Stanley's inequality for any $k \gg 0$. In fact, we proved the following result.

**Theorem 8** ([23], Theorem 2.2 and Corollary 2.6). *Let G be a graph and suppose p is the number of bipartite connected components of G. Then, for every integer $k \geq 0$, we have* $\mathrm{sdepth}(I(G)^k/I(G)^{k+1}) \geq p$. *In particular, $I(G)^k/I(G)^{k+1}$ satisfies Stanley's inequality, for every integer $k \gg 0$.*

We mention that in the special case, when $G$ is a forest, Theorem 8 was proved in ([21] Theorem 3.1).

The *diameter* of a connected graph is the maximum distance between any two vertices. Here, the *distance* between two vertices is the minimum length of a path connecting the vertices.

Fouli and Morey [20] studied the Stanley depth of small powers of edge ideals and determined a lower bound for it.

**Theorem 9** ([20], Theorem 4.18). *Assume that G is a graph with c connected components, and let d denote the maximum of the diameters of the connected components of G. Then, for every integer $1 \leq t \leq 3$, we have*

$$\mathrm{sdepth}(S/I(G)^t) \geq \left\lceil \frac{d - 4t + 5}{3} \right\rceil + c - 1.$$

Fouli and Morey ([20] Corollary 3.3, Theorems 4.4 and 4.13) also show that the inequality of Theorem 9 remains true, if one replaces depth with depth.

## 3. Integral Closure of Powers

The study of Stanley depth of integral closure of powers of monomial ideals was initiated in [28] and continued in [26]. Before stating the results of these papers, we recall some definitions and basic facts from the theory of integral closure.

Let $I \subset S$ be an arbitrary ideal. An element $f \in S$ is *integral* over $I$, if there exists an equation

$$f^k + c_1 f^{k-1} + \ldots + c_{k-1} f + c_k = 0 \quad \text{with } c_i \in I^i.$$

The *integral closure* of $I$, denoted by $\overline{I}$, is the set of elements of $S$ which are integral over $I$. It is known that the integral closure of a monomial ideal $I$ is again a monomial ideal, and it is generated by all monomials $u \in S$ with the property that there exists an integer $k$, such that $u^k \in I^k$ (see ([13] Theorem 1.4.2)). An ideal is said to be *integrally closed* if it is equal to its integral closure, and it is *normal* if all its powers are integrally closed. By ([29] Theorem 3.3.18), a monomial ideal $I$ is normal if, and only if, the Rees algebra $\mathcal{R}(I)$ is a normal ring.

We first notice that there is no general inequality between the Stanley depth of $S/I$ and that of $S/\overline{I}$. This will be illustrated in the following examples.

**Example 2** ([28], Example 1.2). *Let $I = (x_1^2, x_2^2, x_1 x_2 x_3)$ be a monomial ideal in the polynomial ring $S = \mathbb{K}[x_1, x_2, x_3]$. It is not difficult to see that $\overline{I} = (x_1^2, x_2^2, x_1 x_2)$. Then, the maximal ideal $\mathfrak{m} = (x_1, x_2, x_3)$ is an associated prime of $S/I$, and it follows from ([30] Proposition 1.3) that $\mathrm{sdepth}(S/I) = 0$. Since $\mathfrak{m}$ is not an associated prime of $S/\overline{I}$, it follows from ([31] Proposition 2.13) that $\mathrm{sdepth}(S/\overline{I}) \geq 1$. Thus, in this example, $\mathrm{sdepth}(S/I) < \mathrm{sdepth}(S/\overline{I})$.*

**Example 3** ([28], Example 1.3). *Let $I = (x_1^2 x_2^2, x_1^2 x_3^2, x_2^2 x_3^2)$ be a monomial ideal in the polynomial ring $S = \mathbb{K}[x_1, x_2, x_3]$. The maximal ideal $\mathfrak{m} = (x_1, x_2, x_3)$ is not an associated prime of $S/I$ and hence, using ([31] Proposition 2.13), we conclude that $\mathrm{sdepth}(S/I) \geq 1$. On the other hand, using ([32] Theorem 2.4), we know that the maximal ideal $\mathfrak{m}$ is an associated prime of $S/\overline{I}$. Hence, it follows from ([30] Proposition 1.3) that $\mathrm{sdepth}(S/\overline{I}) = 0$. Therefore, in this example, $\mathrm{sdepth}(S/I) > \mathrm{sdepth}(S/\overline{I})$.*

Although there is no general inequality between $\mathrm{sdepth}(S/I)$ and $\mathrm{sdepth}(S/\overline{I})$, we will see in the following theorem that the Stanley depth of $S/\overline{I}$ provides an upper bound for the Stanley depth of the quotient ring of some powers of $I$.

**Theorem 10** ([28], Theorem 2.8). *Let $I_2 \subseteq I_1$ be two monomial ideals in $S$. Then, there exists an integer $k \geq 1$, such that for every $s \geq 1$,*

$$\mathrm{sdepth}(I_1^{sk}/I_2^{sk}) \leq \mathrm{sdepth}(\overline{I_1}/\overline{I_2}).$$

In particular, we have the following corollary.

**Corollary 1.** *Let $I \subset S$ be a monomial ideal. Then, there exist integers $k_1, k_2 \geq 1$, such that for every $s \geq 1$,*

$$\mathrm{sdepth}(I^{sk_1}) \leq \mathrm{sdepth}(\overline{I})$$

*and*

$$\mathrm{sdepth}(S/I^{sk_2}) \leq \mathrm{sdepth}(S/\overline{I}).$$

We mention that the assertions of Corollary 1 remain true if one replaces sdepth with depth, ([26] Theorem 4.5).

In Question 1, we asked whether the high powers of an ideal satisfied Stanley's inequality. One can ask a similar question by replacing $I^k$ with its integral closure. This question is posed in [26].

**Question 3** ([26], Question 1.2). *Let $I$ be a monomial ideal. Is it true that $\overline{I^k}$ and $S/\overline{I^k}$ satisfy Stanley's inequality for every integer $k \gg 0$?*

Before we focus on the above question, we recall the following result of Hoa and Trung concerning the depth of integral closure of high powers of monomial ideals.

**Theorem 11** ([33], Lemma 1.5). *Let $I$ be a monomial ideal of $S$. Then, $\mathrm{depth}(S/\overline{I^k}) = n - \ell(I)$ for every integer $k \gg 0$.*

According to the above theorem, Question 3 is equivalent to the following question.

**Question 4** ([26], Question 1.3). *Let $I$ be a monomial ideal. Is it true that the inequalities $\mathrm{sdepth}(\overline{I^k}) \geq n - \ell(I) + 1$ and $\mathrm{sdepth}(S/\overline{I^k}) \geq n - \ell(I)$ hold, for every integer $k \gg 0$?*

Let $I$ be a monomial ideal of $S$ and assume that $\mathrm{sdepth}(S/I^k) \geq n - \ell(I)$ (resp. $\mathrm{sdepth}(I^k) \geq n - \ell(I) + 1$), for every integer $k \gg 0$. It follows from Corollary 1 that $\mathrm{sdepth}(S/\overline{I^k}) \geq n - \ell(I)$ (resp. $\mathrm{sdepth}(\overline{I^k}) \geq n - \ell(I) + 1$), for every integer $k \gg 0$. Thus, the answers of Questions 3 and 4 are positive for $I$. This argument, together with Theorem 2, implies the following result concerning the Stanley depth of integral closure of powers for the complete intersection of monomial ideals.

**Theorem 12.** *Let $I$ be a complete intersection monomial ideal which is minimally generated by $t$ monomials.*

(i) *For every integer $k \geq 1$, we have*
$$\mathrm{sdepth}(\overline{I^k}) \geq n - t + 1.$$

(ii) *For every integer $k \geq 1$, we have*
$$\mathrm{sdepth}(S/\overline{I^k}) = n - t.$$

Note that in part (ii) of the above theorem, we used the fact that for any complete intersection monomial ideal and any integer $k \geq 1$, the dimension of $S/\overline{I^k}$ is $n - t$, where $t$ is the number of minimal monomial generators of $I$.

Restricting to edge ideals, combining the above argument with Theorems 4 and 7 implies the following results.

**Theorem 13** ([26], Theorem 3.2). *Let $G$ be a graph, and suppose that $p$ is the number of bipartite connected components of $G$. Then, for every integer $k \geq 1$, we have $\mathrm{sdepth}(S/\overline{I(G)^k}) \geq p$. In particular, $S/\overline{I(G)^k}$ satisfies Stanley's inequality for every integer $k \gg 0$.*

**Theorem 14** ([26], Theorem 3.3). *Let $G$ be a non-bipartite graph, and suppose that $p$ is the number of bipartite connected components of $G$. Then, for every integer $k \geq 1$, we have $\mathrm{sdepth}(\overline{I(G)^k}) \geq p + 1$. In particular, $\overline{I(G)^k}$ satisfies Stanley's inequality for every integer $k \gg 0$.*

Assume that $G$ is a bipartite graph. We know from ([13] Theorem 1.4.6 and Corollary 10.3.17) that for any integer $k \geq 1$, the equality $I(G)^k = \overline{I(G)^k}$ holds. Therefore, $\overline{I(G)^k}$ satisfies Stanley's inequality if, and only if $I(G)^k$ satisfies that inequality. Because of this reason, we exclude the case of bipartite graphs in Theorem 14.

Let $I$ be a monomial ideal. It is also reasonable to study the depth and the Stanley depth of $\overline{I^k}/\overline{I^{k+1}}$. In [26], we proved the following result about the depth of these modules for large $k$.

**Theorem 15** ([26], Theorem 4.1). *For any nonzero monomial ideal $I \subsetneq S$, the sequence $\{\mathrm{depth}(\overline{I^k}/\overline{I^{k+1}})\}_{k=0}^{\infty}$ is convergent, and moreover,*
$$\lim_{k \to \infty} \mathrm{depth}(\overline{I^k}/\overline{I^{k+1}}) = n - \ell(I).$$

According to Theorem 15, in order to prove that $\overline{I^k}/\overline{I^{k+1}}$ satisfies Stanley's inequality, for $k \gg 0$, we must show that $\mathrm{sdepth}(\overline{I^k}/\overline{I^{k+1}}) \geq n - \ell(I)$, for high $k$.

Let $I$ be a monomial ideal of $S$ with $\mathrm{sdepth}(I^k/I^{k+1}) \geq n - \ell(I)$ for every integer $k \gg 0$, say for $k \geq k_0$. We fix an integer $k \geq 1$. By Corollary 1, there exists an integer $s$ with $sk \geq k_0$ such that

$$\mathrm{sdepth}(\overline{I^k}/\overline{I^{k+1}}) \geq \mathrm{sdepth}(I^{sk}/I^{s(k+1)}).$$

On the other hand, as $\mathbb{K}$-vector spaces, we have

$$I^{sk}/I^{s(k+1)} = \bigoplus_{i=sk}^{sk+s-1} I^i/I^{i+1}.$$

By the definition of Stanley depth, we conclude that

$$\mathrm{sdepth}(I^{sk}/I^{s(k+1)}) \geq \min\left\{\mathrm{sdepth}(I^i/I^{i+1}) \mid i = sk, \ldots, sk+s-1\right\} \geq n - \ell(I),$$

where the last inequality follows from the assumption. Therefore,

$$\mathrm{sdepth}(\overline{I^k}/\overline{I^{k+1}}) \geq n - \ell(I).$$

Hence, $\overline{I^k}/\overline{I^{k+1}}$ satisfies Stanley's inequality for $k \gg 0$. In particular cases, it follows from Theorems 2 and 8 that $\overline{I^k}/\overline{I^{k+1}}$ satisfies Stanley's inequality, for every integer $k \gg 0$, if $I$ is either a complete intersection monomial ideal or an edge ideal.

Let $I$ be a normal ideal. By ([13] Proposition 10.3.2),

$$\lim_{k \to \infty} \mathrm{depth}(S/I^k) = n - \ell(I).$$

Hence, if $I^k$ and $S/I^k$ satisfy Stanley's inequality for large $k$, we must have

$$\mathrm{sdepth}(S/I^k) \geq n - \ell(I) \quad \text{and} \quad \mathrm{sdepth}(I^k) \geq n - \ell(I) + 1.$$

In fact, in [28], we conjectured that the above inequalities hold in a more general setting.

**Conjecture 3** ([28], Conjecture 2.6). *Let $I \subset S$ be an integrally closed monomial ideal. Then, $\mathrm{sdepth}(S/I) \geq n - \ell(I)$ and $\mathrm{sdepth}(I) \geq n - \ell(I) + 1$.*

The following example shows that the inequalities of Conjecture 3 do not necessarily hold if $I$ is not integrally closed.

**Example 4** ([28], Example 2.5). *Consider the ideal $I = (x_1^2, x_2^2, x_1 x_2 x_3, x_1 x_2 x_4)$ in the polynomial ring $S = \mathbb{K}[x_1, x_2, x_3, x_4]$. Then, $\ell(I) = 2$. However, $\mathfrak{m} = (x_1, x_2, x_3, x_4)$ is an associated prime of $S/I$ and therefore, we conclude from ([30] Proposition 1.3) that $\mathrm{sdepth}(S/I) = 0$ and by ([34] Corollary 1.2), $\mathrm{sdepth}(I) \leq 2$. This shows that the inequalities $\mathrm{sdepth}(S/I) \geq n - \ell(I)$ and $\mathrm{sdepth}(I) \geq n - \ell(I) + 1$ do not hold for $I$.*

As we mentioned in Section 2, the inequalities of Conjecture 3 are true for any polymatroidal ideal (we know from ([19] Theorem 3.4) that any polymatroidal ideal is integrally closed). Also, in ([35] Corollary 3.4), we verified Conjecture 3 for any squarefree monomial ideal which is generated in a single degree.

We close this section by the following result which permits us to compare the Stanley depth of integral closure of a monomial ideal and its powers.

**Theorem 16** ([28], Theorem 2.8). *Let $J \subseteq I$ be two monomial ideals in $S$. Then, for every integer $k \geq 1$,*

$$\mathrm{sdepth}(\overline{I^k}/\overline{J^k}) \leq \mathrm{sdepth}(\overline{I}/\overline{J}).$$

The following corollary is an immediate consequence of Theorem 16.

**Corollary 2.** *Let $I \subset S$ be a monomial ideal. Then, for every integer $k \geq 1$,*

$$\operatorname{sdepth}(\overline{I^k}) \leq \operatorname{sdepth}(\overline{I})$$

*and*

$$\operatorname{sdepth}(S/\overline{I^k}) \leq \operatorname{sdepth}(S/\overline{I}).$$

We mention that the inequalities of Corollary 2 remain true if one replaces sdepth with depth, and this has been proved by Hoa and Trung ([33] Lemma 2.5).

## 4. Symbolic Powers

In this section, we collect the recent results concerning the Stanley depth of symbolic powers of squarefree monomial ideals. We first recall the definition of symbolic powers, and then we continue in two subsections.

**Definition 4.** *Let $I$ be an ideal of $S$, and let $\operatorname{Min}(I)$ denote the set of minimal primes of $I$. For every integer $k \geq 1$, the k-th symbolic power of $I$, denoted by $I^{(k)}$, is defined to be*

$$I^{(k)} = \bigcap_{\mathfrak{p} \in \operatorname{Min}(I)} \operatorname{Ker}(S \to (S/I^k)_{\mathfrak{p}}).$$

Let $I$ be a squarefree monomial ideal in $S$, and suppose that $I$ has the primary decomposition

$$I = \mathfrak{p}_1 \cap \ldots \cap \mathfrak{p}_r,$$

where each $\mathfrak{p}_i$ is a prime ideal generated by a subset of the variables of $S$. It follows from ([13] Proposition 1.4.4) that for every integer $k \geq 1$,

$$I^{(k)} = \mathfrak{p}_1^k \cap \ldots \cap \mathfrak{p}_r^k.$$

### 4.1. Asymptotic Behavior of Stanley Depth of Symbolic Powers

Let $I$ be a squarefree monomial ideal. As we mentioned in Section 2, based on the limit behavior of depth of powers of $I$, Herzog [6] conjectured that the Stanley depth of $S/I^k$ is constant for large $k$ (see Conjecture 1). On the other hand, it is known that if one replaces the ordinary powers by symbolic powers, then again the depth function stabilizes. In fact, Hoa, Kimura, Terai, and Trung [36] are even able to compute the limit value of this function. In order to state their result, we need the following definition.

**Definition 5.** *Suppose $I$ is a squarefree monomial ideal, and let $\mathcal{R}_s(I) = \bigoplus_{k=0}^{\infty} I^{(k)}$ be the symbolic Rees ring of $I$. The Krull dimension of $\mathcal{R}_s(I)/\mathfrak{m}\mathcal{R}(I)$ is called the symbolic analytic spread of $I$ and is denoted by $\ell_s(I)$.*

Let $I$ be a squarefree monomial ideal. Varbaro ([37] Proposition 2.4) showed that

$$\min_k \operatorname{depth}(S/I^{(k)}) = n - \ell_s(I).$$

In [36], Hoa, Kimura, Terai, and Trung proved that the minimum and the limit of the sequence $\{\operatorname{depth}(S/I^{(k)})\}_{k=1}^{\infty}$ coincide. Indeed, they showed the following stronger result. In the following theorem, $\operatorname{bight}(I)$ denotes the maximum height of associated primes of $I$.

**Theorem 17** ([36], Theorem 2.4). *Let I be a squarefree monomial ideal of S. Then,* $\operatorname{depth}(S/I^{(k)}) = n - \ell_s(I)$, *for every integer* $k \geq n(n+1)\operatorname{bight}(I)^{n/2}$.

As the depth function of symbolic powers of a squarefree monomial ideal is eventually constant, one may ask whether the same is true for the Stanley depth—or in other words, whether an analogue of Conjecture 1 is true, if one replaces the ordinary power with a symbolic power. In [38], we gave a positive answer to this question. In fact, we have something more—first, we will see in the following theorem that one can compare the Stanley depth of certain symbolic powers of a squarefree monomial ideal.

**Theorem 18** ([38], Theorem 4.2). *Let $I \subset S$ be a squarefree monomial ideal. Suppose that m and k are positive integers. Then, for every integer j with $m - k \leq j \leq m$, we have*

$$\operatorname{sdepth}(I^{(m)}) \geq \operatorname{sdepth}(I^{(km+j)}) \quad \text{and} \quad \operatorname{sdepth}(S/I^{(m)}) \geq \operatorname{sdepth}(S/I^{(km+j)}).$$

We recall that in the special case of $j = m$, the inequalities of Theorem 18 were also proved in ([39] Corollary 3.2). We also mention that the assertions of Theorem 18 are true if one replaces sdepth with depth, and this was proved independently by Nguyen and Trung ([40] Theorem 2.7), Montaño and Núñez-Betancourt ([41] Theorem 3.4), and the author ([38] Theorem 3.3).

As an immediate consequence of Theorem 18, we obtained the following result.

**Corollary 3** ([38], Corollary 4.3). *For every squarefree monomial ideal $I \subset S$, we have*

$$\operatorname{sdepth}(S/I) \geq \operatorname{sdepth}(S/I^{(2)}) \geq \operatorname{sdepth}(S/I^{(3)})$$

*and*

$$\operatorname{sdepth}(I) \geq \operatorname{sdepth}(I^{(2)}) \geq \operatorname{sdepth}(I^{(3)}).$$

Assume that $I$ is a squarefree monomial ideal, and set

$$m := \min_k \operatorname{sdepth}(S/I^{(k)}).$$

Let $t \geq 1$ be the smallest integer with $\operatorname{sdepth}(S/I^{(t)}) = m$. If $t = 1$, then by Theorem 18, for every integer $k \geq 1$, we have $\operatorname{sdepth}(S/I^{(k)}) = m$. Now, suppose $t \geq 2$. Again, by Theorem 18, we have $\operatorname{sdepth}(S/I^{(t^2-t)}) = m$. For every integer $k > t^2 - t$, we write $k = st + j$, where $s$ and $j$ are positive integers and $1 \leq j \leq t$. As $k > t^2 - t$, we conclude that $s \geq t - 1$. It then follows from Theorem 18 that

$$\operatorname{sdepth}(S/I^{(k)}) = \operatorname{sdepth}(S/I^{(st+j)}) \leq \operatorname{sdepth}(S/I^{(t)}) = m.$$

By the choice of $m$, we conclude that for every integer $k \geq t^2 - t$, the equality $\operatorname{sdepth}(S/I^{(k)}) = m$ holds. Therefore, the sequence $\{\operatorname{sdepth}(S/I^{(k)})\}_{k=1}^{\infty}$ is convergent and

$$\min_k \operatorname{sdepth}(S/I^{(k)}) = m = \lim_{k \to \infty} \operatorname{sdepth}(S/I^{(k)}).$$

Similarly, one proves that the sequence $\{\operatorname{sdepth}(I^{(k)})\}_{k=1}^{\infty}$ is convergent and

$$\min_k \operatorname{sdepth}(I^{(k)}) = \lim_{k \to \infty} \operatorname{sdepth}(I^{(k)}).$$

Therefore, we have the following result.

**Theorem 19** ([38], Theorem 4.4). *For every squarefree monomial ideal $I$, the sequences $\{\text{sdepth}(S/I^{(k)})\}_{k=1}^{\infty}$ and $\{\text{sdepth}(I^{(k)})\}_{k=1}^{\infty}$ are convergent. Moreover,*

$$\min_k \text{sdepth}(S/I^{(k)}) = \lim_{k \to \infty} \text{sdepth}(S/I^{(k)}),$$

*and*

$$\min_k \text{sdepth}(I^{(k)}) = \lim_{k \to \infty} \text{sdepth}(I^{(k)}).$$

A squarefree monomial ideal $I$ is called *normally torsionfree*, if $I^{(k)} = I^k$, for every integer $k \geq 1$. It is immediate from Theorem 19 that for any normally torsionfree squarefree monomial ideal $I$, the sequences $\{\text{sdepth}(S/I^k)\}_{k=1}^{\infty}$ and $\{\text{sdepth}(I^k)\}_{k=1}^{\infty}$ are convergent. In particular, Conjecture 1 is true for normally torsionfree squarefree monomial ideals.

Let $I$ be a squarefree monomial ideal. The smallest integer $t \geq 1$, such that $\text{depth}(S/I^m) = \lim_{k \to \infty} \text{depth}(S/I^k)$ for all $m \geq t$, is called the *index of depth stability of powers* of $I$, and is denoted by $\text{dstab}(I)$. Similarly, one can define the *index of depth stability of symbolic powers* by replacing the ordinary powers with symbolic powers. The index of depth stability of symbolic powers is denoted by $\text{dstab}_s(I)$. By Theorem 17, we have

$$\text{dstab}_s(I) \leq n(n+1)\text{bight}(I)^{n/2}.$$

According to Theorem 19, one can also define the *indices of sdepth stability of symbolic powers*, that is,

$$\text{sdstab}_s(I) = \min\{t \mid \text{sdepth}(I^{(m)}) = \lim_{k \to \infty} \text{sdepth}(I^{(k)}) \text{ for all } m \geq t\}$$

$$\text{sdstab}_s(S/I) = \min\{t \mid \text{sdepth}(S/I^{(m)}) = \lim_{k \to \infty} \text{sdepth}(S/I^{(k)}) \text{ for all } m \geq t\}.$$

We also defined the following quantities:

$$\text{sdmin}_s(I) = \min\{t \mid \text{sdepth}(I^{(t)}) = \lim_{k \to \infty} \text{sdepth}(I^{(k)})\}$$

$$\text{sdmin}_s(S/I) = \min\{t \mid \text{sdepth}(S/I^{(t)}) = \lim_{k \to \infty} \text{sdepth}(S/I^{(k)})\}.$$

The argument before Theorem 19 also proves the following proposition.

**Proposition 1** ([38], Corollary 4.5). *For every squarefree monomial ideal $I \subset S$, we have*

$$\text{sdstab}_s(I) \leq \max\{1, \text{sdmin}_s(I)^2 - \text{sdmin}_s(I)\}$$

*and*

$$\text{sdstab}_s(S/I) \leq \max\{1, \text{sdmin}_s(S/I)^2 - \text{sdmin}_s(S/I)\}.$$

As we mentioned above, the assertions of Theorem 18 are true also for the depth. Thus, a similar argument, as we explained before Theorem 19, implies that the inequalities of Proposition 1 remain true, if one replaces Stanley depth with depth. This has been already observed in ([38] Theorem 3.6).

Let $I$ be a squarefree monomial ideal. We know from Theorem 19 that the sequences $\{\text{sdepth}(S/I^{(k)})\}_{k=1}^{\infty}$ and $\{\text{sdepth}(I^{(k)})\}_{k=1}^{\infty}$ are convergent. Now, it is natural to ask the following question.

**Question 5.** *Let $I$ be a squarefree monomial ideal. How can one describe the limits of the sequences $\{\text{sdepth}(S/I^{(k)})\}_{k=1}^{\infty}$ and $\{\text{sdepth}(I^{(k)})\}_{k=1}^{\infty}$?*

Question 5 is widely open. We know the answer only for very special classes of ideals. For example, assume that $I$ is a squarefree complete intersection monomial ideal. It is easy to check that for any integer $k \geq 1$, the equality $I^{(k)} = I^k$ holds. Therefore, using Theorem 2, we conclude that

$$\lim_{k \to \infty} \mathrm{sdepth}(S/I^{(k)}) = n - t,$$

and

$$\lim_{k \to \infty} \mathrm{sdepth}(I^{(k)}) = n - t + 1,$$

where $t$ is the number of minimal monomial generators of $I$ (which is also equal to $\ell_s(I)$).

We are also able to compute the limit of the sequence $\{\mathrm{sdepth}(S/I^{(k)})\}_{k=1}^{\infty}$, where $I$ is the Stanley-Reisner ideal of a matroid. We first recall some basic definitions from the theory of Stanley-Reisner rings.

A *simplicial complex* $\Delta$ on the set of vertices $V(\Delta) = [n] := \{1, \ldots, n\}$ is a collection of subsets of $[n]$ which contains $\{i\}$ for any $i \in [n]$, and is closed under taking subsets; that is, if $F \in \Delta$ and $F' \subseteq F$, then also $F' \in \Delta$. Every element $F \in \Delta$ is called a *face* of $\Delta$. The *dimension* of a face $F$ is defined to be $|F| - 1$. The *dimension* of $\Delta$ which is denoted by $\dim \Delta$, is defined to be $d - 1$, where $d = \max\{|F| \mid F \in \Delta\}$. The *Stanley-Reisner ideal* of $\Delta$ is defined as

$$I_\Delta = (\prod_{i \in F} x_i : F \subseteq [n], F \notin \Delta) \subseteq S.$$

**Definition 6.** *A simplicial complex $\Delta$ is called matroid if, for every pair of faces $F, G \in \Delta$ with $|F| > |G|$, there is a vertex $x \in F \setminus G$ such that $G \cup \{x\}$ is a face of $\Delta$.*

As we mentioned above, there is some information about the limit of the Stanley depth function of symbolic powers of the Stanley-Reisner ideal of a matroid.

**Theorem 20** ([38], Theorem 4.7). *Let $\Delta$ be a matroid. Then,*

$$\lim_{k \to \infty} \mathrm{sdepth}(S/I_\Delta^{(k)}) = n - \ell_s(I_\Delta) = \dim \Delta + 1$$

*and*

$$\lim_{k \to \infty} \mathrm{sdepth}(I_\Delta^{(k)}) \geq n - \ell_s(I_\Delta) + 1.$$

4.2. Cover Ideals

Let $G$ be a graph with vertex set $V(G) = \{x_1, \ldots, x_n\}$. A subset $C$ of $V(G)$ is called a *vertex cover* of $G$ if every edge of $G$ is incident to at least one vertex of $C$. A vertex cover $C$ is called a *minimal vertex cover* of $G$ if no proper subset of $C$ is a vertex cover of $G$. The *cover ideal* of $G$ is a squarefree monomial ideal of $S$ which is defined as

$$J(G) = (\prod_{x_i \in C} x_i \mid C \text{ is a minimal vertex cover of } G).$$

It is easy to see that the cover ideal is the Alexander dual of the edge ideal, that is,

$$J(G) = I(G)^{\vee} = \bigcap_{\{x_i, x_j\} \in E(G)} (x_1, x_j).$$

Let $I$ be a squarefree monomial ideal. In Question 1, we asked whether $I^k$ and $S/I^k$ satisfied Stanley's inequality for every integer $k \gg 0$. One can also ask the similar question for symbolic powers.

**Question 6** ([7], Question 1.2). *Let I be a monomial ideal. Is it true that $I^{(k)}$ and $S/I^{(k)}$ satisfy Stanley's inequality for every integer $k \gg 0$?*

In this subsection, we investigate the above question for cover ideals. By Theorem 17, in order to know whether the high symbolic powers of cover ideals satisfy Stanley's inequality, we need to compute their symbolic analytic spread. This has been done by Constantinescu and Varbaro [42]. Indeed, they provide a combinatorial description for the symbolic analytic spread of $J(G)$. To state their result, we need to recall some notions from the graph theory.

Let $G$ be a graph. A *matching* in $G$ is a set of edges such that no two different edges share a common vertex. A subset $W$ of $V(G)$ is called an *independent subset* of $G$ if there are no edges among the vertices of $W$. Let $M = \{\{a_i, b_i\} \mid 1 \leq i \leq r\}$ be a nonempty matching of $G$. We say that $M$ is an *ordered matching* of $G$ if the following conditions hold.

(1) $A := \{a_1, \ldots, a_r\}$ is an independent subset of vertices of $G$, and
(2) $\{a_i, b_j\} \in E(G)$ implies that $i \leq j$.

The *ordered matching number* of $G$, denoted by $\nu_o(G)$, is defined to be

$$\nu_o(G) = \max\{|M| \mid M \subseteq E(G) \text{ is an ordered matching of } G\}.$$

**Theorem 21** ([42], Theorem 2.8). *For any graph $G$,*

$$\ell_s(J(G)) = \nu_o(G) + 1.$$

As a consequence of Theorems 17 and 21, for any graph $G$ with $n$ vertices, we have

$$\lim_{k \to \infty} \text{depth}(S/J(G)^{(k)}) = n - \nu_o(G) - 1.$$

Hoa, Kimura, Terai, and Trung [36], determined a linear upper bound for the index of depth stability of symbolic powers of cover ideals. In [7], we provided an alternative proof for their result.

**Theorem 22** ([36], Theorem 3.4 and [7], Theorem 3.1). *Let $G$ be a graph with $n$ vertices. Then, for every integer $k \geq 2\nu_o(G) - 1$, we have*

$$\text{depth}(S/J(G)^{(k)}) = n - \nu_o(G) - 1.$$

In [7], we also proved that high symbolic powers of cover ideals satisfy Stanley's inequality. Indeed, we proved the following result.

**Theorem 23** ([7], Theorem 3.5 and Corollary 3.6). *Let $G$ be a graph with $n$ vertices. Then, for every integer $k \geq 1$, we have*

$$\text{sdepth}(J(G)^{(k)}) \geq n - \nu_o(G) \quad \text{and} \quad \text{sdepth}(S/J(G)^{(k)}) \geq n - \nu_o(G) - 1.$$

*In particular, $J(G)^{(k)}$ and $S/J(G)^{(k)}$ satisfy the Stanley's inequality, for every integer $k \geq 2\nu_o(G) - 1$.*

The assertions of Theorem 23 for the special case of bipartite graphs was also proved in [43].

Let $G$ be a graph with $n$ vertices. We say $G$ is very well-covered if $n$ is an even integer and moreover, every vertex cover of $G$ has size $n/2$. The graph $G$ is called Cohen-Macaulay if the ring $S/I(G)$ is Cohen-Macaulay. We know from Theorem 23 that for any graph $G$, the modules $J(G)^{(k)}$ and $S/J(G)^{(k)}$ satisfy Stanley's inequality for $k \gg 0$. However, in the case of Cohen-Macaulay very well-covered graphs, we have something more.

**Proposition 2** ([44], Corollary 3.8). *Let G be a Cohen-Macaulay very well-covered graph. Then, $J(G)^{(k)}$ and $S/J(G)^{(k)}$ satisfy Stanley's inequality for every integer $k \geq 1$.*

In Question 5, we asked about the limit values of the sequences $\{\mathrm{sdepth}(S/I^{(k)})\}_{k=1}^{\infty}$ and $\{\mathrm{sdepth}(I^{(k)})\}_{k=1}^{\infty}$, where $I$ is a squarefree monomial ideal. For the case of cover ideals, we pose the following conjecture.

**Conjecture 4.** *Let G be a graph with n vertices. Then,*

$$\lim_{k \to \infty} \mathrm{sdepth}(S/J(G)^{(k)}) = n - \nu_o(G) - 1,$$

*and*

$$\lim_{k \to \infty} \mathrm{depth}(J(G)^{(k)}) = n - \nu_o(G).$$

Let $I$ be a squarefree monomial ideal. According to Theorem 17, the sequence $\{\mathrm{depth}(S/I^{(k)})\}_{k=1}^{\infty}$ is convergent. The situation is even better if $I$ is a cover ideal. In fact, Hoa, Kimura, Terai, and Trung ([36] Theorem 3.2) proved that the above sequence is non-increasing for cover ideals. In other words, for every graph $G$ and any integer $k \geq 1$, we have

$$\mathrm{depth}(S/J(G)^{(k)}) \geq \mathrm{depth}(S/J(G)^{(k+1)}).$$

We recall that the above inequality for bipartite graphs was also proved in ([45] Theorem 3.2).

We close this article by mentioning that the above inequality is true if one replaces depth with sdepth. In fact, we have the following result.

**Theorem 24** ([7], Theorem 3.3). *Let G be a graph. Then, for every integer $k \geq 1$, we have:*

(i) $\mathrm{sdepth}(S/J(G)^{(k)}) \geq \mathrm{sdepth}(S/J(G)^{(k+1)})$, *and*
(ii) $\mathrm{sdepth}(J(G)^{(k)}) \geq \mathrm{sdepth}(J(G)^{(k+1)})$.

**Funding:** This research is partially funded by the Simons Foundation Grant Targeted for Institute of Mathematics, Vietnam Academy of Science and Technology.

**Acknowledgments:** The author is grateful to Siamak Yassemi for encouraging him to write this survey article. The author also thanks the reviewers for careful reading of the paper and for useful comments.

**Conflicts of Interest:** The author declares no conflict of interest.

### References

1. Pournaki, M.R.; Fakhari, S.A.S.; Tousi, M.; Yassemi, S. What is … Stanley depth? *Not. Am. Math. Soc.* **2009**, *56*, 1106–1108.
2. Stanley, R.P. Linear Diophantine equations and local cohomology. *Invent. Math.* **1982**, *68*, 175–193. [CrossRef]
3. Ichim, B.; Katthän, L.; Moyano-Fernández, J.J. The behavior of Stanley depth under polarization. *J. Combin. Theory Ser. A* **2015**, *135*, 332–347. [CrossRef]
4. Duval, A.M.; Goeckner, B.; Klivans, C.J.; Martin, J.L. A non-partitionable Cohen-Macaulay simplicial complex. *Adv. Math.* **2016**, *299*, 381–395. [CrossRef]
5. Herzog, J.; Jahan, A.S.; Yassemi, S. Stanley decompositions and partitionable simplicial complexes. *J. Algebraic Combin.* **2008**, *27*, 113–125. [CrossRef]
6. Herzog, J. A survey on Stanley depth. In *Monomial Ideals, Computations and Applications*; Lecture Notes in Math; Bigatti, A., Giménez, P., Sáenz-de-Cabezón, E., Eds.; Springer: Berlin, Germany, 2013.
7. Fakhari, S.A.S. Depth and Stanley depth of symbolic powers of cover ideals of graphs. *J. Algebra* **2017**, *492*, 402–413. [CrossRef]
8. Biró, C.; Howard, D.M.; Keller, M.T.; Trotter, W.T.; Young, S.J. Interval partitions and Stanley depth. *J. Combin. Theory Ser. A* **2010**, *117*, 475–482. [CrossRef]

9. Cimpoeaş, M. Some remarks on the Stanley depth for multigraded modules. *Le Math.* **2008**, *LXIII*, 165–171.
10. Cimpoeaş, M. On the Stanley depth of powers of some classes of monomial ideals. *Bull. Iran. Math. Soc.* **2018**, *44*, 739–747. [CrossRef]
11. Mohammadi, F.; Moradi, S. Weakly polymatroidal ideals with applications to vertex cover ideals. *Osaka J. Math.* **2010**, *47*, 627–636.
12. Fakhari, S.A.S. Stanley depth of weakly polymatroidal ideals. *Arch. Math.* **2014**, *103*, 229–233. [CrossRef]
13. Herzog, J.; Hibi, T. *Monomial Ideals*; Springer: Berlin, Germany, 2011.
14. Herzog, J.; Takayama, Y. Resolutions by mapping cones, in: The Roos Festschrift volume Nr.2(2). *Homol. Homotopy Appl.* **2002**, *4*, 277–294. [CrossRef]
15. Jahan, A.S. Prime filtrations and Stanley decompositions of squarefree modules and Alexander duality. *Manuscr. Math.* **2009**, *130*, 533–550. [CrossRef]
16. Pournaki, M.R.; Fakhari, S.A.S.; Yassemi, S. On the Stanley depth of weakly polymatroidal ideals. *Arch. Math.* **2013**, *100*, 115–121. [CrossRef]
17. Burch, L. Codimension and analytic spread. *Proc. Camb. Philos. Soc.* **1972**, *72*, 369–373. [CrossRef]
18. Brodmann, M. The asymptotic nature of the analytic spread. *Math. Proc. Camb. Philos. Soc.* **1979**, *86*, 35–39. [CrossRef]
19. Herzog, J.; Rauf, A.; Vladoiu, M. The stable set of associated prime ideals of a polymatroidal ideal. *J. Algebraic Combin.* **2013**, *37*, 289–312. [CrossRef]
20. Fouli, L.; Morey, S. A lower bound for depths of powers of edge ideals. *J. Algebraic Combin.* **2015**, *42*, 829–848. [CrossRef]
21. Alipour, A.; Fakhari, S.S.; Yassemi, S. Stanley depth of factor of polymatroidal ideals and edge ideal of forests. *Arch. Math.* **2015**, *105*, 323–332. [CrossRef]
22. Pournaki, M.R.; Fakhari, S.A.S.; Yassemi, S. Stanley depth of powers of the edge ideal of a forest. *Proc. Am. Math. Soc.* **2013**, *141*, 3327–3336. [CrossRef]
23. Fakhari, S.A.S. On the Stanley depth of powers of edge ideals. *J. Algebra* **2017**, *489*, 463–474. [CrossRef]
24. Trung, T.N. Stability of depth of power of edge ideals. *J. Algebra* **2016**, *452*, 157–187. [CrossRef]
25. Vasconcelos, W. *Integral Closure, Rees Algebras, Multiplicities, Algorithms*; Springer Monographs in Mathematics; Springer: Berlin, Germany, 2005.
26. Fakhari, S.A.S. On the depth and Stanley depth of the integral closure of powers of monomial ideals. *Collect. Math.* **2018**, 1–13. Available online: https://arxiv.org/abs/1808.03189 (accessed on 7 June 2019).
27. Herzog, J.; Hibi, T. The depth of powers of an ideal. *J. Algebra* **2005**, *291*, 325–650. [CrossRef]
28. Fakhari, S.A.S. Stanley depth of the integral closure of monomial ideals. *Collect. Math.* **2013**, *64*, 351–362. [CrossRef]
29. Villarreal, R.H. *Monomial Algebras*; Dekker: New York, NY, USA, 2001.
30. Herzog, J.; Vladoiu, M.; Zheng, X. How to compute the Stanley depth of a monomial ideal. *J. Algebra* **2009**, *322*, 3151–3169. [CrossRef]
31. Bruns, W.; Krattenthaler, C.; Uliczka, J. Stanley decompositions and Hilbert depth in the Koszul complex. *J. Commut. Algebra* **2010**, *2*, 327–357. [CrossRef]
32. Jarrah, A.S. Integral closures of Cohen–Macaulay monomial ideals. *Commun. Algebra* **2002**, *30*, 5473–5478. [CrossRef]
33. Hoa, L.T.; Trung, T.N. Stability of depth and Cohen-Macaulayness of integral closures of powers of monomial ideals. *Acta Math. Vietnam.* **2018**, *43*, 67–81. [CrossRef]
34. Ishaq, M. Values and bounds of the Staney depth. *Carpathian J. Math.* **2011**, *27*, 217–224.
35. Fakhari, S.A.S. Stanley depth of weakly polymatroidal ideals and squarefree monomial ideals. *Ill. J. Math.* **2013**, *57*, 871–881. [CrossRef]
36. Hoa, L.T.; Kimura, K.; Terai, N.; Trung, T.N. Stability of depths of symbolic powers of Stanley-Reisner ideal. *J. Algebra* **2017**, *473* 307–323. [CrossRef]
37. Varbaro, M. Symbolic powers and matroids. *Proc. Am. Math. Soc.* **2011**, *139*, 2357–2366. [CrossRef]
38. Fakhari, S.A.S. Stability of depth and Stanley depth of symbolic powers of squarefree monomial ideals. *arXiv* **2018**, arXiv:1812.03742.
39. Fakhari, S.A.S. Stanley depth and symbolic powers of monomial ideals. *Math. Scand.* **2017**, *120*, 5–16. [CrossRef]
40. Nguyen, H.D.; Trung, N.V. Depth functions of symbolic powers of homogeneous ideals. **2018**, preprint.

41. Montaño, J.; Núñez-Betancourt, L. Splittings and symbolic powers of square-free monomial ideals. *arXiv* **2018**, arXiv:1809.02308.
42. Constantinescu, A.; Varbaro, M. Koszulness, Krull dimension, and other properties of graph-related algebras. *J. Algebraic Combin.* **2011**, *34*, 375–400. [CrossRef]
43. Fakhari, S.A.S. Depth, Stanley depth and regularity of ideals associated to graphs. *Arch. Math.* **2016**, *107*, 461–471. [CrossRef]
44. Fakhari, S.A.S. Symbolic powers of cover ideal of very well-covered and bipartite graphs. *Proc. Am. Math. Soc.* **2018**, *146*, 97–110. [CrossRef]
45. Constantinescu, A.; Pournaki, M.R.; Seyed Fakhari, S.A.; Terai, N.; Yassemi, S. Cohen-Macaulayness and limit behavior of depth for powers of cover ideals. *Comm. Algebra* **2015**, *43*, 143–157. [CrossRef]

© 2019 by the authors. Licensee MDPI, Basel, Switzerland. This article is an open access article distributed under the terms and conditions of the Creative Commons Attribution (CC BY) license (http://creativecommons.org/licenses/by/4.0/).

*Article*

# The Regularity of Some Families of Circulant Graphs

Miguel Eduardo Uribe-Paczka [1] and Adam Van Tuyl [2,*]

[1] Departamento de Matemáticas, Escuela Superior de Física y Matemáticas, Instituto Politécnico Nacional, Mexico City 07300, Mexico
[2] Department of Mathematics and Statistics, McMaster University, Hamilton, ON L8S 4L8, Canada
* Correspondence: vantuyl@math.mcmaster.ca; Tel.: +1-905-525-9140 (ext. 27016)

Received: 18 June 2019; Accepted: 18 July 2019; Published: 22 July 2019

**Abstract:** We compute the Castelnuovo–Mumford regularity of the edge ideals of two families of circulant graphs, which includes all cubic circulant graphs. A feature of our approach is to combine bounds on the regularity, the projective dimension, and the reduced Euler characteristic to derive an exact value for the regularity.

**Keywords:** circulant graphs; edge ideals; Castelnuovo–Mumford regularity; projective dimension

**MSC:** 13D02; 05C25; 13F55

---

## 1. Introduction

Let $G$ be any finite simple graph with vertex set $V(G) = \{x_1, \ldots, x_n\}$ and edge set $E(G)$, where simple means no loops or multiple edges. The *edge ideal* of $G$ is the ideal $I(G) = \langle x_i x_j \mid \{x_i, x_j\} \in E(G) \rangle$ in $R = k[x_1, \ldots, x_n]$, a standard graded polynomial ring over a field $k$ ($k$ is any field). Describing the dictionary between the graph theoretic properties of $G$ and the algebraic properties of $I(G)$ or $R/I(G)$ is an active area of research; e.g., see [1,2].

Relating the homological invariants of $I(G)$ and the graph theoretic invariants of $G$ has proven to be a fruitful approach to building this dictionary. Recall that the *minimal graded free resolution* of $I(G) \subseteq R$ is a long exact sequence of the form:

$$0 \to \bigoplus_j R(-j)^{\beta_{l,j}(I(G))} \to \bigoplus_j R(-j)^{\beta_{l-1,j}(I(G))} \to \cdots \to \bigoplus_j R(-j)^{\beta_{0,j}(I(G))} \to I(G) \to 0$$

where $l \leq n$. Here, $R(-j)$ denotes the free $R$-module obtained by shifting the degrees of $R$ by $j$, that is $R(-j)_a = R_{a-j}$. We denote by $\beta_{i,j}(I(G))$ the $i,j^{\text{th}}$ *graded Betti number* of $I(G)$; this number equals the number of minimal generators of degree $j$ in the $i^{\text{th}}$ syzygy module of $I(G)$. Two invariants that measure the "size" of the resolution are the *(Castelnuovo–Mumford) regularity* and the *projective dimension*, defined as:

$$\operatorname{reg}(I(G)) = \max\{j - i \mid \beta_{i,j}(I(G)) \neq 0\}, \text{ and}$$
$$\operatorname{pd}(I(G)) = \max\{i \mid \beta_{i,j}(I(G)) \neq 0 \text{ for some } j\}.$$

One wishes to relate the numbers $\beta_{i,j}(I(G))$ to the invariants of $G$; e.g., see the survey of Hà [3], which focuses on describing $\operatorname{reg}(I(G))$ in terms of the invariants of $G$.

In this note, we give explicit formulas for $\operatorname{reg}(I(G))$ for the edge ideals of two infinite families of circulant graphs. Our results complement previous work on the algebraic and combinatorial topological properties of circulant graphs (e.g, [4–11]). Fix an integer $n \geq 2$ and a subset $S \subseteq \{1, \ldots, \lfloor \frac{n}{2} \rfloor\}$. The *circulant graph* $C_n(S)$ is the graph on the vertex set $\{x_1, \ldots, x_n\}$ such that $\{x_i, x_j\} \in E(C_n(S))$

if and only if $|i-j|$ or $n - |i-j| \in S$. To simplify notation, $C_n(a_1, \ldots, a_t)$ is sometimes written for $C_n(\{a_1, \ldots, a_t\})$. As an example, the graph $C_{10}(1,3)$ is drawn in Figure 1.

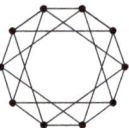

**Figure 1.** The circulant $C_{10}(1,3)$.

When $S = \{1, \ldots, \lfloor \frac{n}{2} \rfloor\}$, then $C_n(S) \cong K_n$, the clique on $n$ vertices. On the other hand, if $S = \{1\}$, then $C_n(1) \cong C_n$, the cycle on $n$ vertices. For both of these families, the regularity of their edge ideals is known. Specifically, the ideal $I(K_n)$ has a linear resolution by Fröberg's theorem [12], so reg($I(K_n)$) = 2. The value of reg($I(C_n)$) can be deduced from the work of Jacques ([13], Theorem 7.6.28). One can view these circulant graphs as "extremal" cases in the sense that $|S|$ is either as large or as small as possible.

Our motivation is to understand the next open cases. In particular, generalizing the case of $K_n$, we compute reg($I(C_n(S))$) when $S = \{1, \ldots, \hat{j}, \ldots, \lfloor \frac{n}{2} \rfloor\}$ for any $1 \leq j \leq \lfloor \frac{n}{2} \rfloor$ (Theorem 5). For most $j$, the regularity follows from Fröberg's theorem and a result of Nevo [14]. To generalize the case of $C_n$ (a circulant graph where every vertex has degree two), we compute the regularity of the edge ideal of any cubic (every vertex has degree three) circulant graph, that is $G = C_{2n}(a,n)$ with $1 \leq a \leq n$ (Theorem 8). Our proof of Theorem 8 requires a new technique to compute reg($I$) for a square-free monomial ideal. Specifically, we show how to use partial information about reg($I$), pd($I$), and the reduced Euler characteristic of the simplicial complex associated with $I$ to determine reg($I$) exactly (see Theorem 4). We believe this result to be of independent interest.

We use the following outline. We first recall the relevant background regarding graph theory and commutative algebra, along with our new result on the regularity of square-free monomial ideals. In Section 3, we compute the regularity of $I(G)$ for the family of graphs $G = C_n(1, \ldots, \hat{j}, \ldots, \lfloor \frac{n}{2} \rfloor)$. In Section 4, we give an explicit formula for the regularity of edge ideals of cubic circulant graphs.

## 2. Background

We review the relevant background from graph theory and commutative algebra. In addition, we give a new result on the regularity of square-free monomial ideals.

*2.1. Graph Theory Preliminaries*

Let $G = (V(G), E(G))$ denote a finite simple graph. We abuse notation and write $xy$ for the edge $\{x, y\} \in E(G)$. The *complement* of $G$, denoted $G^c$, is the graph $(V(G^c), E(G^c))$ where $V(G^c) = V(G)$ and $E(G^c) = \{xy \mid xy \notin E(G)\}$. The *neighbours* of $x \in V(G)$ are the set $N(x) = \{y \in V(G) \mid xy \in E(G)\}$. The *closed neighbourhood* of $x$ is $N[x] = N(x) \cup \{x\}$. The *degree* of $x$ is $\deg(x) = |N(x)|$. If we need to highlight the graph, we write $N_G[x]$ or $N_G(x)$.

A graph $H = (V(H), E(H))$ is a *subgraph* of $G$ if $V(H) \subseteq V(G)$ and $E(H) \subseteq E(G)$. Given a subset $W \subseteq V(G)$, the *induced subgraph* of $G$ on $W$ is the graph $G_W = (W, E(G_W))$ where $E(G_W) = \{xy \in E(G) \mid \{x, y\} \subseteq W\}$. Notice that an induced subgraph is a subgraph of $G$, but not every subgraph of $G$ is an induced subgraph.

An *n-cycle*, denoted $C_n$, is the graph with $V(C_n) = \{x_1, \ldots, x_n\}$ and edges $E(C_n) = \{x_1x_2, x_2x_3, \ldots, x_{n-1}x_n, x_nx_1\}$. A graph $G$ has a *cycle* of length $n$ if $G$ has a subgraph of the form $C_n$. A graph is a *chordal graph* if $G$ has no induced graph of the form $C_n$ with $n \geq 4$. A graph $G$ is *co-chordal* if $G^c$ is chordal. The *co-chordal number* of $G$, denoted co-chord($G$), is the smallest number of subgraphs of $G$ such that $G = G_1 \cup \cdots \cup G_s$, and each $G_i^c$ is a chordal graph.

A *claw* is the graph with $V(G) = \{x_1, x_2, x_3, x_4\}$ with edges $E(G) = \{x_1x_2, x_1x_3, x_1x_4\}$. A graph is *claw-free* if no induced subgraph of the graph is a claw. A graph $G$ is *gap-free* if no induced subgraph of $G^c$ is a $C_4$. Finally, the *complete graph* $K_n$ is the graph with $V(K_n) = \{x_1, \ldots, x_n\}$ and $E(K_n) = \{x_i x_j \mid 1 \leq i < j \leq n\}$.

*2.2. Algebraic Preliminaries*

We recall some facts about the regularity of $I(G)$. Note that for any homogeneous ideal, $\text{reg}(I) = \text{reg}(R/I) + 1$.

We collect together a number of useful results on the regularity of edge ideals.

**Theorem 1.** *Let $G$ be a finite simple graph. Then:*

(i) *if $G = H \cup K$, with $H$ and $K$ disjoint, then:*

$$\text{reg}(R/I(G)) = \text{reg}(R/I(H)) + \text{reg}(R/I(K)).$$

(ii) $\text{reg}(I(G)) = 2$ *if and only if $G^c$ is a chordal graph.*
(iii) $\text{reg}(I(G)) \leq \text{co-chord}(G) + 1$.
(iv) *if $G$ is gap-free and claw-free, then $\text{reg}(I(G)) \leq 3$.*
(v) *if $x \in V(G)$, then $\text{reg}(I(G)) \in \{\text{reg}(I(G \setminus N_G[x])) + 1, \text{reg}(I(G \setminus x))\}$.*

**Proof.** For (i), see Woodroofe ([15], Lemma 8). Statement (ii) is Fröberg's Theorem ([12], Theorem 1). Woodroofe ([15], Theorem 1) first proved (iii). Nevo first proved (iv) in [14] (Theorem 5.1). For (v), see Dao, Huneke, and Schweig ([16], Lemma 3.1). □

We require a result of Kalai and Meshulam [17] that has been specialized to edge ideals.

**Theorem 2.** *([17], Theorems 1.4 and 1.5) Let $G$ be a finite simple graph, and suppose $H$ and $K$ are subgraphs such that $G = H \cup K$. Then,*

(i) $\text{reg}(R/I(G)) \leq \text{reg}(R/I(H)) + \text{reg}(R/I(K))$, *and*
(ii) $\text{pd}(I(G)) \leq \text{pd}(I(H)) + \text{pd}(I(K)) + 1$.

We now introduce a new result on the regularity of edge ideals. In fact, because our result holds for all square-free monomial ideals, we present the more general case.

We review some facts about simplicial complexes. Given a vertex set $V = \{x_1, \ldots, x_n\}$, a simplicial complex $\Delta$ on $V$ is a set of subsets of $V$ that satisfies the properties: (i) if $F \in \Delta$ and $G \subseteq F$, then $G \in \Delta$, and (ii) $\{x_i\} \in \Delta$ for $i = 1, \ldots, n$. Note that $\emptyset \in \Delta$ by (i) since $\{x_1\} \in \Delta$ by (ii) (if $\Delta$ is not the empty complex). An element of $\Delta$ is called a *face*. For any $W \subseteq V$, the restriction of $\Delta$ to $W$ is the simplicial complex $\Delta_W = \{F \in \Delta \mid F \subseteq W\}$.

The *dimension* of $F \in \Delta$ is $\dim(F) = |F| - 1$. The *dimension* of a complex $\Delta$, denoted $\dim(\Delta)$, is $\max\{\dim(F) \mid F \in \Delta\}$. Let $f_i$ equal the number of faces of $\Delta$ of dimension $i$; we adopt the convention that $f_{-1} = 1$. If $\dim(\Delta) = D$, then the *f-vector* of $\Delta$ is the $(D+2)$-tuple $f(\Delta) = (f_{-1}, f_0, \ldots, f_D)$.

We can associate with any simplicial complex $\Delta$ on $V$ a monomial ideal $I_\Delta$ in the polynomial ring $R = k[x_1, \ldots, x_n]$ (with $k$ a field) as follows:

$$I_\Delta = \langle x_{j_1} x_{j_2} \cdots x_{j_r} \mid \{x_{j_1}, x_{j_2}, \ldots, x_{j_r}\} \notin \Delta \rangle.$$

The ideal $I_\Delta$ is the *Stanley–Reisner ideal* of $\Delta$. This construction can be reversed. Given a square-free monomial ideal $I$ of $R$, the simplicial complex associated with $I$ is:

$$\Delta(I) = \{\{x_{i_1}, \ldots, x_{i_r}\} \mid \text{the square-free monomial } x_{i_1} \cdots x_{i_r} \notin I\}.$$

Given a square-free monomial ideal $I$, Hochster's formula relates the Betti numbers of $I$ to the reduced simplicial homology of $\Delta(I)$. See [2] (Section 6.2) for more background on $\tilde{H}_j(\Gamma;k)$, the $j^{\text{th}}$ reduced simplicial homology group of a simplicial complex $\Gamma$.

**Theorem 3.** (Hochster's formula) *Let $I \subseteq R = k[x_1,\ldots,x_n]$ be a square-free monomial ideal, and set $\Delta = \Delta(I)$. Then, for all $i,j \geq 0$,*

$$\beta_{i,j}(I) = \sum_{|W|=j,\ W \subseteq V} \dim_k \tilde{H}_{j-i-2}(\Delta_W;k).$$

Given a simplicial complex $\Delta$ of dimension $D$, the dimensions of the homology groups $\tilde{H}_i(\Delta;k)$ are related to the $f$-vector $f(\Delta)$ via the *reduced Euler characteristic*:

$$\tilde{\chi}(\Delta) = \sum_{i=-1}^{D} (-1)^i \dim_k \tilde{H}_i(\Delta;k) = \sum_{i=-1}^{D} (-1)^i f_i. \tag{1}$$

Note that the reduced Euler characteristic is normally defined to be equal to one of the two sums, and then, one proves the two sums are equal (e.g., see [2], Section 6.2).

Our new result on the regularity of square-free monomial ideals allows us to determine $\text{reg}(I)$ exactly if we have enough partial information on the regularity, projective dimension, and the reduced Euler characteristic.

**Theorem 4.** *Let $I$ be a square-free monomial ideal of $R = k[x_1,\ldots,x_n]$ with associated simplicial complex $\Delta = \Delta(I)$.*

(i) *Suppose that $\text{reg}(I) \leq r$ and $\text{pd}(I) \leq n - r + 1$.*

  (a) *If $r$ is even and $\tilde{\chi}(\Delta) > 0$, then $\text{reg}(I) = r$.*
  (b) *If $r$ is odd and $\tilde{\chi}(\Delta) < 0$, then $\text{reg}(I) = r$.*

(ii) *Suppose that $\text{reg}(I) \leq r$ and $\text{pd}(I) \leq n - r$. If $\tilde{\chi}(\Delta) \neq 0$, then $\text{reg}(I) = r$.*

**Proof.** By Hochster's formula (Theorem 3), note that $\beta_{a,n}(I) = \dim_k \tilde{H}_{n-a-2}(\Delta;k)$ for all $a \geq 0$ since the only subset $W \subseteq V$ with $|W| = n$ is $V$.

(i) If $\text{reg}(I) \leq r$ and $\text{pd}(I) \leq n - r + 1$, we have $\beta_{a,n}(I) = 0$ for all $a \leq n - r - 1$ and $\beta_{a,n}(I) = 0$ for all $a \geq n - r + 2$. Consequently, among all the graded Betti numbers of the form $\beta_{a,n}(I)$ as $a$ varies, only $\beta_{n-r,n}(I) = \dim_k \tilde{H}_{r-2}(\Delta;k)$ and $\beta_{n-r+1,n}(I) = \dim_k \tilde{H}_{r-3}(\Delta;k)$ may be non-zero. Thus, by (1):

$$\tilde{\chi}(\Delta) = (-1)^{r-2} \dim_k \tilde{H}_{r-2}(\Delta;k) + (-1)^{r-3} \dim_k \tilde{H}_{r-3}(\Delta;k).$$

If we now suppose that $r$ is even and $\tilde{\chi}(\Delta) > 0$, the above expression implies:

$$\dim_k \tilde{H}_{r-2}(\Delta;k) - \dim_k \tilde{H}_{r-3}(\Delta;k) > 0,$$

and thus, $\beta_{n-r,n}(I) = \dim_k \tilde{H}_{r-2}(\Delta;k) \neq 0$. As a consequence, $\text{reg}(I) = r$, thus proving (a). Similarly, if $r$ is odd and $\tilde{\chi}(\Delta) < 0$, this again forces $\beta_{n-r,n}(I) = \dim_k \tilde{H}_{r-2}(\Delta;k) \neq 0$, thus proving (b).

(ii) Similar to Part (i), the hypotheses on the regularity and projective dimension imply that $\tilde{\chi}(\Delta) = (-1)^{r-2} \dim_k \tilde{H}_{r-2}(\Delta;k) = (-1)^{r-2} \beta_{n-r,n}(I)$. Therefore, if $\tilde{\chi}(\Delta) \neq 0$, then $\beta_{n-r,n}(I) \neq 0$, which implies $\text{reg}(I) = r$.

□

**Remark 1.** *There is a similar result to Theorem 4 for the projective dimension of I. In particular, under the assumptions of (i) and if r is even and $\tilde{\chi}(\Delta) < 0$, or if r is odd and $\tilde{\chi}(\Delta) > 0$, then the proof of Theorem 4 shows that $\mathrm{pd}(I) = n - r + 1$. Under the assumptions of (ii), then $\mathrm{pd}(I) = n - r$.*

We will apply Theorem 4 to compute the regularity of cubic circulant graphs (see Theorem 8). We will also require the following terminology and results that relate the reduced Euler characteristic to the independence polynomial of a graph.

A subset $W \subseteq V(G)$ is an *independent set* if for all $e \in E(G)$, $e \not\subseteq W$. The set of independent sets forms a simplicial complex called the *independence complex* of $G$, that is,

$$\mathrm{Ind}(G) = \{W \mid W \text{ is an independent set of } V(G)\}.$$

Note that $\mathrm{Ind}(G) = \Delta_{I(G)}$, the simplicial complex associated with the edge ideal $I(G)$.

The *independence polynomial* of a graph $G$ is defined as:

$$I(G, x) = \sum_{r=0}^{\alpha} i_r x^r,$$

where $i_r$ is the number of independent sets of cardinality $r$. Note that $(i_0, i_1, \ldots, i_\alpha) = (f_{-1}, f_0, \ldots, f_{\alpha-1})$ is the $f$-vector of $\mathrm{Ind}(G)$. Since $\tilde{\chi}(\mathrm{Ind}(G)) = \sum_{i=-1}^{\alpha-1}(-1)^i f_i$, we get:

$$\tilde{\chi}(\mathrm{Ind}(G)) = -I(G, -1). \tag{2}$$

Thus, the value of $\tilde{\chi}(\mathrm{Ind}(G))$ can be extracted from the independence polynomial $I(G, x)$.

## 3. The Regularity of the Edge Ideals of $C_n(1, \ldots, \widehat{j}, \ldots, \lfloor \frac{n}{2} \rfloor)$

In this section, we compute the regularity of the edge ideal of the circulant graph $G = C_n(S)$ with $S = \{1, \ldots, \widehat{j}, \ldots, \lfloor \frac{n}{2} \rfloor\}$ for any $j \in \{1, \ldots, \lfloor \frac{n}{2} \rfloor\}$.

We begin with the observation that the complement of $G$ is also a circulant graph, and in particular, $G^c = C_n(j)$. Furthermore, we have the following structure result.

**Lemma 1.** *Let $H = C_n(j)$ with $1 \leq j \leq \lfloor \frac{n}{2} \rfloor$, and set $d = \gcd(j, n)$. Then, H is the union of d disjoint cycles of length $\frac{n}{d}$. Furthermore, H is a chordal graph if and only if $n = 2j$ or $n = 3j$.*

**Proof.** Label the vertices of $H$ as $\{0, 1, \ldots, n-1\}$, and set $d = \gcd(j, n)$. For each $0 \leq i < d$, the induced graph on the vertices $\{i, j+i, 2j+i, \ldots, (\frac{n}{d}-1)j+i\}$ is a cycle of length $\frac{n}{d}$, thus proving the first statement (if $\frac{n}{d} = 2$, then $H$ consists of $d$ disjoint edges). For the second statement, if $n = 3j$, then $d = \gcd(j, n) = 3$, so $H$ is the disjoint union of three cycles, and thus chordal. If $n = 2j$, then $H$ consists of $j$ disjoint edges and, consequently, is chordal. Otherwise, $\frac{n}{d} \geq 4$, and so, $H$ is not chordal. □

**Lemma 2.** *Let $G = C_n(1, \ldots, \widehat{j}, \ldots, \lfloor \frac{n}{2} \rfloor)$, and $d = \gcd(j, n)$.*

*(i) If $\frac{n}{d} \geq 4$, then G is claw-free.*
*(ii) If $\frac{n}{d} \geq 5$, then G is gap free.*

**Proof.** For the first statement, suppose that $G$ has an induced subgraph $H$ on $\{z_1, z_2, z_3, z_4\} \subseteq V(G)$ that is a claw. Then, $H^c$ is an induced subgraph of $G^c$ of the form:

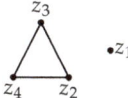

119

However, by Lemma 1, the induced cycles of $G^c$ have length $\frac{n}{d} \geq 4$. Thus, $G$ is claw-free.

The second statement also follows from Lemma 1 and the fact that a graph $G$ is gap-free if and only if $G^c$ has no induced four cycles. □

The main result of this section is given below.

**Theorem 5.** *If $G = C_n(1, \ldots, \widehat{j}, \ldots, \lfloor \frac{n}{2} \rfloor)$, then:*

$$\mathrm{reg}(I(G)) = \begin{cases} 2 & n = 2j \text{ or } n = 3j \\ 3 & \text{otherwise.} \end{cases}$$

**Proof.** Consider $G^c = C_n(j)$, and let $d = \gcd(j, n)$. By Lemma 1, $G^c$ consists of induced cycles of length $k = \frac{n}{d}$. Because $1 \leq j \leq \lfloor \frac{n}{2} \rfloor$, we have $d < n$, and thus, $2 \leq k \leq n$. If $k = 2$ or 3, i.e., if $n = 2j$ or $n = 3j$, Lemma 1 and Theorem 1 (ii) combine to give $\mathrm{reg}(I(G)) = 2$. If $k \geq 5$, then Lemmas 1 and 2 imply that $G$ is gap-free and claw-free (but not chordal), and so, Theorem 1 (ii) and (iv) implies $\mathrm{reg}(I(G)) = 3$.

To compete the proof, we need to consider the case $k = 4$. In this case, $n = 4d$, and so, $G = C_{4d}(1, \ldots, \widehat{j}, \ldots, 2d)$. However, because $d = \gcd(j, 4d)$ and $1 \leq j \leq 2d$, we have $d = j$. Therefore, the graph $G$ has the form $G = C_{4j}(1, \ldots, \widehat{j}, \ldots, 2j)$. By Lemma 1, $G^c$ is $j$ disjoint copies of $C_4$, and thus, Theorem 1 (ii) gives $\mathrm{reg}(I(G)) \geq 3$. To prove that $\mathrm{reg}(I(G)) = 3$, we show co-chord$(G) = 2$ and apply Theorem 1 (iii).

Label the vertices of $G$ as $0, 1, \ldots, 4j - 1$, and let:

$$V_1 = \{0, 1, 2, \ldots, j-1, 2j, 2j+1, \ldots, 3j-1\} \text{ and}$$
$$V_2 = \{j, j+1, \ldots, 2j-1, 3j, 3j+1, \ldots, 4j-1\}.$$

Observe that the induced subgraph of $G$ on $V_1$ (and $V_2$) is the complete graph $K_{2j}$.

Let $G_1$ be the graph with $V(G_1) = V(G)$ and edge set $E(G_1) = (E(C_{4j}(1, \ldots, j-1)) \cup E(G_{V_1})) \setminus E(G_{V_2})$. Similarly, we let $G_2$ be the graph with $V(G_2) = V(G)$ and edge set $E(G_2) = (E(C_{4j}(j+1, \ldots, 2j)) \cup E(G_{V_2})) \setminus E(G_{V_1})$.

We now claim that $G = G_1 \cup G_2$, and furthermore, both $G_1^c$ and $G_2^c$ are chordal and, consequently, co-chord$(G) = 2$. The equality $G = G_1 \cup G_2$ follows from the fact that:

$$E(G_1) \cup E(G_2) = E(C_{4j}(1, \ldots, j-1)) \cup E(C_{4j}(j+1, \ldots, 2j))$$
$$= E(G_{4j}(1, \ldots, \widehat{j}, \ldots, 2j)).$$

To show that $G_1^c$ is chordal, first note that the induced graph on $V_1$, that is $(G_1)_{V_1}$, is the complete graph $K_{2j}$. In addition, the vertices $V_2$ form an independent set of $G_1$. To see why, note that if $a, b \in V_2$ are such that $ab \in E(G)$, then $ab \in E(G_{V_2})$. However, by the construction of $E(G_1)$, none of the edges of $E(G_{V_2})$ belong to $E(G_1)$. Therefore, $ab \notin E(G_1)$, and thus, $V_2$ is an independent set in $G_1$.

The above observations therefore imply that in $G_1^c$, the vertices of $V_1$ form an independent set, and $(G_1^c)_{V_2}$ is the clique $K_{2j}$. To show that $G_1^c$ is chordal, suppose that $G_1^c$ has an induced cycle of length $t \geq 4$ on $\{v_1, v_2, v_3, \ldots, v_t\}$. Since the induced graph on $(G_1^c)_{V_2}$ is a clique, at most two of the vertices of $\{v_1, v_2, \ldots, v_t\}$ can belong to $V_2$. Indeed, if there were at least three $v_i, v_j, v_k \in \{v_1, v_2, \ldots, v_t\} \cap V_2$, then the induced graph on these vertices is a three cycle, contradicting the fact that $\{v_1, v_2, \ldots, v_t\}$ is the minimal induced cycle of length $t \geq 4$. However, then at least $t - 2 \geq 2$ vertices of $\{v_1, v_2, \ldots, v_t\}$ must belong to $V_1$, and in particular, at least two of them are adjacent. However, this cannot happen since the vertices of $V_1$ are independent in $G_1^c$. Thus, $G_1^c$ must be a chordal graph.

The proof that $G_2^c$ is chordal is similar. Note that the vertices of $V_2$ are an independent set, and $(G_2^c)_{V_1}$ is the clique $K_{2j}$. The proof now proceeds as above. □

## 4. Cubic Circulant Graphs

We now compute the regularity of the edge ideals of *cubic circulant graphs*, that is a circulant graph where every vertex has degree three. In general, a circulant graph $C_n(a_1, \ldots, a_t)$ is $2t$-regular (every vertex has degree $2t$), except if $2a_t = n$, in which case, it is $(2t-1)$-regular. Consequently, cubic circulant graphs have the form $G = C_{2n}(a, n)$ with integers $1 \leq a \leq n$. The main result of this section can also be viewed as an application of Theorem 4 to compute the regularity of a square-free monomial ideal.

We begin with a structural result for cubic circulants due to Davis and Domke.

**Theorem 6.** *[18] Let $1 \leq a < n$ and $t = \gcd(2n, a)$.*

(a) *If $\frac{2n}{t}$ is even, then $C_{2n}(a, n)$ is isomorphic to $t$ copies of $C_{\frac{2n}{t}}(1, \frac{n}{t})$.*
(b) *If $\frac{2n}{t}$ is odd, then $C_{2n}(a, n)$ is isomorphic to $\frac{t}{2}$ copies of $C_{\frac{4n}{t}}(2, \frac{2n}{t})$.*

Theorem 6 implies that a cubic circulant graph is the disjoint union of one or more connected cubic circulant graphs. Furthermore, the only connected cubic circulant graphs are those circulant graphs that are isomorphic to either the circulant $C_{2n}(1, n)$ for any $n \geq 2$ or the circulant $C_{2n}(2, n)$ with $n > 1$ odd (for the second circulant, if $n$ is not odd, then Theorem 6 implies that this circulant is not connected). Recall from Theorem 1 $(i)$ that, to compute the regularity of a graph, it is enough to compute the regularity of each connected component. Therefore, it suffices to compute the regularity of the edge ideals of $C_{2n}(1, n)$ and $C_{2n}(2, n)$ with $n$ odd. Moving forward, unless stated otherwise, we will restrict to connected cubic circulant graphs. Note it will be convenient to use the representation and labelling of these two graphs as in Figure 2.

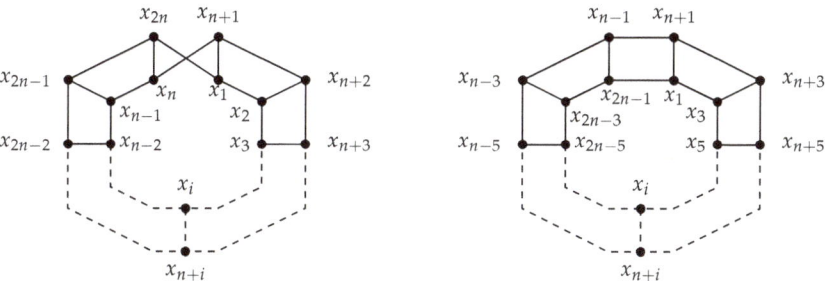

**Figure 2.** The graphs $C_{2n}(1, n)$ and $C_{2n}(2, n)$.

Our strategy is to use Theorem 4 to compute the regularity of these two graphs. Thus, we need bounds on $\mathrm{reg}(I(G))$ and $\mathrm{pd}(I(G))$ and information about the reduced Euler characteristic of $\mathrm{Ind}(G)$ when $G = C_{2n}(1, n)$ or $C_{2n}(2, n)$.

We first bound the regularity and the projective dimension. We introduce the following three families of graphs, where the $t \geq 1$ denotes the number of "squares":

(i) The family $A_t$:

(ii) The family $B_t$:

(iii) The family $D_t$:

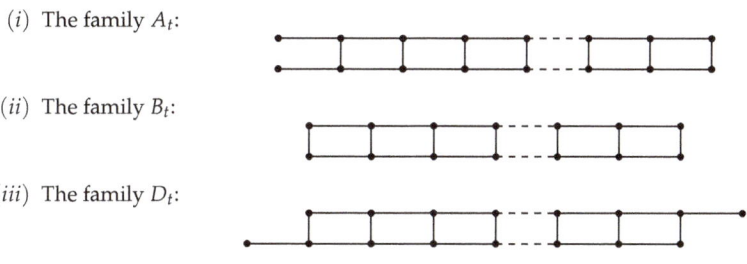

**Lemma 3.** *With the notation as above, we have:*

(i) *If* $G = A_t$, *then:*

$$\operatorname{reg}(I(G)) \leq \begin{cases} \frac{t+4}{2} & \text{if } t \text{ even} \\ \frac{t+3}{2} & \text{if } t \text{ odd} \end{cases} \quad \text{and} \quad \operatorname{pd}(I(G)) \leq \begin{cases} \frac{3t}{2} + 1 & \text{if } t \text{ even} \\ \frac{3(t-1)}{2} + 2 & \text{if } t \text{ odd}. \end{cases}$$

(ii) *If* $G = B_t$, *then*

$$\operatorname{reg}(I(G)) \leq \begin{cases} \frac{t+4}{2} & \text{if } t \text{ even} \\ \frac{t+3}{2} & \text{if } t \text{ odd}. \end{cases}$$

(iii) *If* $G = D_t$ *and* $t = 2l + 1$ *with* $l$ *an odd number, then* $\operatorname{reg}(I(G)) \leq \frac{t+3}{2}$.

**Proof.** (i) The proof is by induction on $t$. Via a direct computation (for example, using *Macaulay2* [19]), one finds $\operatorname{reg}(I(A_1)) = 2$, $\operatorname{reg}(I(A_2)) = 3$, $\operatorname{pd}(I(A_1)) = 2$, and $\operatorname{pd}(I(A_2)) = 4$. Our values agree with the upper bounds given in the statement, so the base cases hold.

Now, suppose that $t \geq 3$. The graph $A_t$ can be decomposed into the subgraphs $A_1$ and $A_{t-2}$, i.e.,

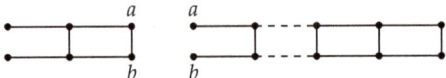

Suppose that $t$ is even. By Theorem 2 and by induction (and the fact that $\operatorname{reg}(R/I) = \operatorname{reg}(I) - 1$), we get:

$$\operatorname{reg}(R/I(A_t)) \leq \operatorname{reg}(R/I(A_1)) + \operatorname{reg}(R/I(A_{t-2})) \leq 1 + \frac{(t-2)+4}{2} - 1 = \frac{t+4}{2} - 1$$

and:

$$\operatorname{pd}(I(A_t)) \leq \operatorname{pd}(I(A_1)) + \operatorname{pd}(I(A_{t-2})) + 1 \leq 2 + \frac{3(t-2)}{2} + 1 + 1 = \frac{3t}{2} + 1.$$

Because the proof for when $t$ is odd is similar, we omit it.

(ii) A direct computation shows $\operatorname{reg}(I(B_1)) = 2$ and $\operatorname{reg}(I(B_2)) = 3$. If $t \geq 3$, we decompose $B_t$ into the subgraphs $B_1$ and $A_{t-2}$, i.e.,

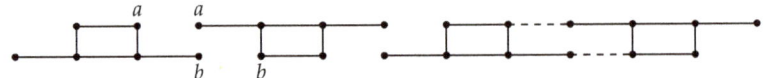

Suppose that $t$ is even. Since $\operatorname{reg}(I(B_1)) = 2$, Theorem 2 and Part (i) above give us:

$$\operatorname{reg}(R/I(B_t)) \leq \operatorname{reg}(R/I(B_1)) + \operatorname{reg}(R/I(A_{t-2})) \leq \frac{(t-2)+4}{2} = \frac{t+2}{2}.$$

Therefore, $\operatorname{reg}(I(B_t)) \leq \frac{t+2}{2} + 1 = \frac{t+4}{2}$. When $t$ is odd, the proof is similar.

(iii) Because $t = 2l + 1$ with $l$ odd, the graph $D_t$ can be decomposed into $l + 1$ subgraphs of the form $A_1$, i.e.,

Since $\operatorname{reg}(I(A_1)) = 2$, by Theorem 2, we get $\operatorname{reg}(R/I(D_t)) \leq (l+1)\operatorname{reg}(R/I(A_1)) = l+1$. Thus, $\operatorname{reg}(I(D_t)) \leq l+2 = \frac{t+3}{2}$. □

**Remark 2.** *In the above proof, we relied on computer computations for our base case. In general, the graded Betti numbers of an ideal may depend on the characteristic of the ground field. However, as shown by Katzman [20], the Betti numbers of edge ideals of graphs on 11 or less vertices are independent of the characteristic. Since the graphs in our induction steps have 11 or less vertices, the values found for our base cases hold in all characteristics.*

We now bound the projective dimensions of the edge ideals of $C_{2n}(1,n)$ and $C_{2n}(2,n)$. In the next two lemmas, we assume that $n \geq 4$. However, as we show in Theorem 8, the bounds (in fact, they are equalities) given in these lemmas also hold if $n = 2$ or $3$, i.e., if $G = C_4(1,2), C_6(1,3)$ or $C_6(2,3)$.

**Lemma 4.** *Let $n \geq 4$.*

(i) *If $G = C_{2n}(1,n)$, then:*
$$\operatorname{pd}(I(G)) \leq \begin{cases} 3k-1 & \text{if } n = 2k \\ 3k+1 & \text{if } n = 2k+1. \end{cases}$$

(ii) *If $G = C_{2n}(2,n)$, then $\operatorname{pd}(I(G)) \leq 3k+1$ where $n = 2k+1$.*

**Proof.** (i) Let $G = C_{2n}(1,n)$, and suppose that $n = 2k+1$. The graph $C_{2n}(1,n)$ can be decomposed into the subgraphs $A_1$ and $A_{2k-2}$, i.e.,

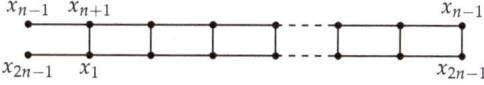

Note that since $n \geq 4$ and $n$ is odd, $2k - 2 \geq 2$. Combining Theorem 2 and Lemma 3, we get:

$$\operatorname{pd}(I(C_{2n}(1,n))) \leq \operatorname{pd}(I(A_{2k-2})) + \operatorname{pd}(I(A_1)) + 1 \leq \left(\frac{3(2k-2)}{2} + 1\right) + 3 = 3k+1.$$

If $n = 2k$, $C_{2n}(1,n)$ can be decomposed as in the previous case with the only difference being that $C_{2n}(1,n)$ can be decomposed into the union of the subgraphs $A_1$ and $A_{2k-3}$. By Theorem 2 and Lemma 3:

$$\operatorname{pd}(I(C_{2n}(1,n))) \leq \operatorname{pd}(I(A_{2k-3})) + \operatorname{pd}(I(A_1)) + 1 \leq \left(\frac{3(2k-4)}{2} + 2\right) + 3 = 3k-1.$$

(ii) Let $G = C_{2n}(2,n)$ with $n = 2k+1$. We can draw $G$ as:

The previous representation of $G$ contains $2k$ squares. Then, the graph $G$ can be decomposed into the subgraphs $A_1$ and $A_{2k-2}$, and the proof runs as in (i). □

We now determine bounds on the regularity.

**Lemma 5.** *Let $n \geq 4$.*

(i) If $G = C_{2n}(1, n)$, then:

$$\operatorname{reg}(I(G)) \leq \begin{cases} k+1 & \text{if } n = 2k, \text{ or if } n = 2k+1 \text{ and } k \text{ odd} \\ k+2 & \text{if } n = 2k+1 \text{ and } k \text{ even.} \end{cases}$$

(ii) If $G = C_{2n}(2, n)$, then

$$\operatorname{reg}(I(G)) \leq \begin{cases} k+1 & \text{if } n = 2k+1 \text{ and } k \text{ even} \\ k+2 & \text{if } n = 2k+1 \text{ and } k \text{ odd.} \end{cases}$$

**Proof.** (i) Let $G = C_{2n}(1, n)$. We consider three cases.

Case 1. $n = 2k$.

In Lemma 4 (i), we saw that $G$ can be decomposed into the subgraphs $A_1$ and $A_{2k-3}$. By Theorem 2 and Lemma 3, we get:

$$\operatorname{reg}(R/I(G)) \leq \operatorname{reg}(R/I(A_1)) + \operatorname{reg}(R/I(A_{2k-3})) \leq k.$$

Case 2. $n = 2k + 1$ with $k$ an odd number.

Using Theorem 1 (v), we have:

$$\operatorname{reg}(I(G)) \in \{\operatorname{reg}(I(G \setminus x_1), \operatorname{reg}(I(G \setminus N_G[x_1]) + 1\}.$$

If we set $W = G \setminus x_1$, then by applying Theorem 1 (v) again, we have:

$$\operatorname{reg}(I(G)) \in \{\operatorname{reg}(I(W \setminus x_{n+1}), \operatorname{reg}(I(W \setminus N_W[x_{n+1}]) + 1, \operatorname{reg}(I(G \setminus N_G[x_1]) + 1\}.$$

We have $G \setminus N_G[x_1] \cong W \setminus N_W[x_{n+1}] \cong D_{2k-3}$. Moreover, $2k - 3 = 2(k - 2) + 1$, and since $k$ is an odd number, $k - 2$ is also odd. Thus, by Lemma 3 (iii), we obtain $\operatorname{reg}(I(D_{2k-3})) \leq \frac{2k-3+3}{2} = k$. On the other hand, the graph $W \setminus x_{n+1} = (G \setminus x_1) \setminus x_{n+1} \cong B_{2k-1}$, so by Lemma 3 (ii), we have $\operatorname{reg}(I(W \setminus x_{n+1})) \leq \frac{2k-1+3}{2} \leq k+1$. Thus, $\operatorname{reg}(I(G)) \leq k+1$.

Case 3. $n = 2k + 1$ with $k$ an even number.

In Lemma 4 (i), we saw that $G$ can be decomposed into the subgraphs $A_1$ and $A_{2k-2}$, and the proof runs as in Case 1.

(ii) Let $G = C_{2n}(2, n)$. We consider two cases.

Case 1. $n = 2k + 1$ with $k$ an even number.

As in the second case of (i), by Theorem 1 (v), we have:

$$\operatorname{reg}(I(G)) \in \{\operatorname{reg}(I(W \setminus x_{n+1}), \operatorname{reg}(I(W \setminus N_W[x_{n+1}]) + 1, \operatorname{reg}(I(G \setminus N_G[x_1]) + 1\}.$$

where $W = G \setminus x_1$. In particular, $W \setminus N_W[x_{n+1}] \cong G \setminus N_G[x_1]$. The graph $G \setminus N_G[x_1]$ can be represented as:

The previous representation of $G \setminus N_G[x_1]$ contains $2k - 3$ squares. It follows that $G \setminus N_G[x_1]$ can be decomposed into the subgraphs $D_{2k-5}$ and $A_1$, i.e.,

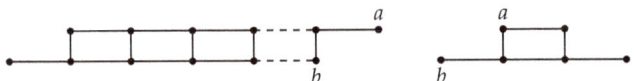

Note that $2k - 5 = 2(k - 3) + 1$, and because $k$ is even, then $k - 3$ is odd. Using Theorem 2 and Lemma 3, we get:

$$\text{reg}(R/I(G \setminus N_G[x_1])) \leq \text{reg}(R/I(D_{2k-5})) + \text{reg}(R/I(A_1)) \leq \frac{2k-2}{2} = k - 1.$$

The graph $W \setminus x_{n+1} \cong B_{2k-1}$. Therefore, by Lemma 3 (ii), we have $\text{reg}(I(W \setminus x_{n+1})) \leq \frac{2k-1+3}{2} = k + 1$. Consequently, $\text{reg}(I(G)) \leq k + 1$, as desired.

Case 2. $n = 2k + 1$ with $k$ an odd number.

The result follows from the fact that the graphs $C_{2n}(2, n)$ can be decomposed into the subgraphs $A_1$ and $A_{2k-2}$ as seen in Lemma 4, and so, $\text{reg}(I(G)) \leq \text{reg}(I(A_1)) + \text{reg}(I(A_{2k-2})) - 1$. □

Our final ingredient is a result of Hoshino ([21], Theorem 2.26) (also see Brown–Hoshino ([22]), Theorems 3.2 and 3.5), which describes the independence polynomial for cubic circulant graphs.

**Theorem 7** ([21,22]). *For each $n \geq 3$, set:*

$$I_n(x) = 1 + \sum_{\ell=0}^{\lfloor \frac{n-2}{4} \rfloor} \frac{2n}{2\ell+1} \binom{n-2\ell-2}{2\ell} x^{2\ell+1}(1+x)^{n-4\ell-2}.$$

(i) *If $G = C_{2n}(1, n)$ with $n$ even, or if $G = C_{2n}(2, n)$ with $n$ odd, then $I(G, x) = I_n(x)$.*
(ii) *If $G = C_{2n}(1, n)$ and $n$ is odd, then $I(G, x) = I_n(x) + 2x^n$.*

We now come to the main result of this section.

**Theorem 8.** *Let $1 \leq a < n$ and $t = \gcd(2n, a)$.*

(a) *If $\frac{2n}{t}$ is even, then:*

$$\text{reg}(I(C_{2n}(a, n))) = \begin{cases} kt + 1 & \text{if } \frac{n}{t} = 2k, \text{ or } \frac{n}{t} = 2k+1 \text{ with } k \text{ an odd number} \\ (k+1)t + 1 & \text{if } \frac{n}{t} = 2k+1 \text{ with } k \text{ an even number.} \end{cases}$$

(b) *If $\frac{2n}{t}$ is odd, then:*

$$\text{reg}(I(C_{2n}(a, n))) = \begin{cases} \frac{kt}{2} + 1 & \text{if } \frac{2n}{t} = 2k+1 \text{ with } k \text{ an even number} \\ \frac{(k+1)t}{2} + 1 & \text{if } \frac{2n}{t} = 2k+1 \text{ with } k \text{ an odd number.} \end{cases}$$

**Proof.** The formulas can be verified directly for the special cases that $n = 2$ (i.e., $G = C_4(1,2)$) or $n = 3$ (i.e., $G = C_6(1,3)$ and $C_6(2,3)$). We can therefore assume $n \geq 4$. In light of Theorem 6 and Theorem 1 (i), it will suffice to prove that the inequalities of Lemma 5 are actually equalities. We will make use of Theorem 4. We consider five cases, where the proof of each case is similar.

Case 1. $G = C_{2n}(1, n)$ with $n = 2k$.

In this case, Lemma 4 gives $\text{pd}(I(G)) \leq 3k - 1$, and Lemma 5 gives $\text{reg}(I(G)) \leq k + 1$. Furthermore, since $\widetilde{\chi}(\text{Ind}(G)) = -I(G, -1)$ by Equation (2), Theorem 7 gives $\widetilde{\chi}(\text{Ind}(G)) = -1$ if $n \neq 4m + 2$, and $\widetilde{\chi}(\text{Ind}(G)) = 3$ if $n = 4m + 2$. Because $G$ has $4k = (k+1) + (3k-1)$ vertices and since $\widetilde{\chi}(\text{Ind}(G)) \neq 0$, Theorem 4 (ii) implies $\text{reg}(I(G)) = k + 1$.

Case 2. $G = C_{2n}(1, n)$ with $n = 2k + 1$ and $k$ even.

We have $\text{reg}(I(G)) \leq k + 2$ and $\text{pd}(I(G)) \leq 3k + 1 = (4k+2) - (k+2) + 1 = n - (k+2) + 1$ by Lemmas 4 and 5, respectively. Because $n$ is odd, $\widetilde{\chi}(\text{Ind}(G)) = -[I_n(-1) + 2(-1)^n] = -[1-2] = 1 > 0$. Therefore, $\text{reg}(I(G)) = k + 2$ by Theorem 4 (i) (a) because $k + 2$ is even and $\widetilde{\chi}(\text{Ind}(G)) = 1 > 0$.

*Case 3.* $G = C_{2n}(1, n)$ with $n = 2k + 1$ and $k$ odd.

We have $\text{reg}(I(G)) = k + 1$ by Theorem 4 (*ii*) because $\text{reg}(I(G)) \leq k + 1$ (Lemma 5), $\text{pd}(I(G)) \leq 3k + 1$ (Lemma 4), $2n = 4k + 2$ is the number of variables, and $\widetilde{\chi}(\text{Ind}(G)) = -1 \neq 0$.

*Case 4.* $G = C_{2n}(2, n)$ with $n = 2k + 1$ and $k$ even.

We have $\text{reg}(I(G)) = k + 1$ from Theorem 4 (*ii*) since $\text{reg}(I(G)) \leq k + 1$ (Lemma 5), $\text{pd}(I(G)) \leq 3k + 1$ (Lemma 4), and $\widetilde{\chi}(\text{Ind}(G)) = -I(G, -1) = -1 \neq 0$ (Theorem 7).

*Case 5.* $G = C_{2n}(2, n)$ with $n = 2k + 1$ and $k$ odd.

In our final case, $\text{reg}(I(G)) \leq k + 2$ by Lemma 5, $\text{pd}(I(G)) \leq 3k + 1$ by Lemma 4. Since $n$ is odd, $\widetilde{\chi}(\text{Ind}(G)) = -I(G, -1) = -1 < 0$ by Theorem 7. Since $k$ is odd, $k + 2$ is odd. Because $2n = 4k + 2$ is the number of variables, we have $\text{reg}(I(G)) = k + 2$ by Theorem 4 (*i*) (*b*).

These five cases now complete the proof. □

**Author Contributions:** Writing, original draft preparation: M.E.U.-P. and A.V.T.; writing, review: M.E.U.-P. and A.V.T.

**Funding:** Van Tuyl's research was funded by NSERC Grant Number RGPIN-2019-05412.

**Acknowledgments:** The authors thank Federico Galetto and Andrew Nicas for their comments and suggestions. We also thank the referees for their helpful comments, suggestions, and corrections. Computations using *Macaulay2* inspired some of our results. The first author thanks CONACYT for financial support.

**Conflicts of Interest:** The authors declare no conflict of interest.

## References

1. Morey, S.; Villarreal, R.H. Edge ideals: Algebraic and combinatorial properties. In *Progress in Commutative Algebra 1*; de Gruyter: Berlin, Gemany, 2012; pp. 85–126.
2. Villarreal, R.H. Monomial algebras. In *Monographs and Research Notes in Mathematics*; CRC Press: Boca Raton, FL, USA, 2015.
3. Hà, H.T. Regularity of squarefree monomial ideals. In *Connections between Algebra, Combinatorics, and Geometry*; Springer Proceedings in Mathematics & Statistics: New York, NY, USA, 2014; Volume 76, pp. 251–276,
4. Earl, J.; Vander Meulen, K.N.; Van Tuyl, A. Independence complexes of well-covered circulant graphs. *Experiment. Math.* **2016**, *25*, 441–451, [CrossRef]
5. Makvand, M.A.; Mousivand, A. Betti numbers of some circulant graphs. To appear *Czechoslov. Math. J.* **2019**, [CrossRef]
6. Mousivand, A. Circulant $S_2$ graphs. *Preprint* **2015**, arXiv:1512.08141.
7. Rinaldo, G. Some algebraic invariants of edge ideal of circulant graphs. *Bull. Math. Soc. Sci. Math. Roumanie (N.S.)* **2018**, *61*, 95–105.
8. Rinaldo, G.; Romeo, F. On the reduced Euler characteristic of independence complexes of circulant graphs. *Discrete Math.* **2018**, *341*, 2380–2386, [CrossRef]
9. Rinaldo, G.; Romeo, F. 2-Dimensional vertex decomposable circulant graphs. *Preprint* **2018**, arXiv:1807.05755.
10. Romeo, F. Chordal circulant graphs and induced matching number. *Preprint* **2018**, arXiv:1811.06409.
11. Vander Meulen, K.N.; Van Tuyl, A.; Watt, C. Cohen-Macaulay Circulant Graphs. *Comm. Alg.* **2014**, *42*, 1896–1910, [CrossRef]
12. Fröberg, R. On Stanley-Reisner rings. In *Topics in Algebra, Part 2 (Warsaw, 1988)*; Banach Center Publ.: Warsaw, Poland, 1990; Volume 2, pp. 57–70.
13. Jacques, S. Betti Numbers of Graph Ideals. Ph.D. Thesis, University of Sheffield, Sheffield, UK, 2004.
14. Nevo, E. Regularity of edge ideals of $C_4$-free graphs via the topology of the lcm-lattice. *J. Combin. Theory Ser. A* **2011**, *118*, 491–501, [CrossRef]
15. Woodroofe, R. Matchings, coverings, and Castelnuovo–Mumford regularity. *J. Commut. Algebra* **2014**, *6*, 287–304, [CrossRef]

16. Dao, H.; Huneke, C.; Schweig, J. Bounds on the regularity and projective dimension of ideals associated to graphs. *J. Algebraic Combin.* **2013**, *38*, 37–55, [CrossRef]
17. Kalai, G.; Meshulam, R. Intersections of Leray complexes and regularity of monomial ideals. *J. Combin. Theory Ser. A* **2006**, *113*, 1586–1592, [CrossRef]
18. Davis, G.J.; Domke, G.S. 3-Circulant Graphs. *J. Combin. Math. Combin. Comput.* **2002**, *40*, 133–142.
19. Grayson, D.; Stillman, M. Macaulay 2, a Software System for Research in Algebraic Geometry. Available online: http://www.math.uiuc.edu/Macaulay2/ (accessed on 20 July 2019).
20. Katzman, M. Characteristic-independence of Betti numbers of graph ideals. *J. Combin. Theory Ser. A* **2006**, *113*, 435–454, [CrossRef]
21. Hoshino, R. Independence Polynomials of Circulant Graphs. Ph.D. Thesis, Dalhousie University, Halifax, NS, Canada, 2008.
22. Brown, J.; Hoshino, R. Well-covered circulant graphs. *Discrete Math.* **2011**, *311*, 244–251. [CrossRef]

© 2019 by the authors. Licensee MDPI, Basel, Switzerland. This article is an open access article distributed under the terms and conditions of the Creative Commons Attribution (CC BY) license (http://creativecommons.org/licenses/by/4.0/).

MDPI
St. Alban-Anlage 66
4052 Basel
Switzerland
Tel. +41 61 683 77 34
Fax +41 61 302 89 18
www.mdpi.com

*Mathematics* Editorial Office
E-mail: mathematics@mdpi.com
www.mdpi.com/journal/mathematics

www.ingramcontent.com/pod-product-compliance
Lightning Source LLC
LaVergne TN
LVHW071957080526
838202LV00064B/6768